Leckie×Leckie
Scotland's leading educational publishers

Success guides

D1081346

HIGHER

Mathematics

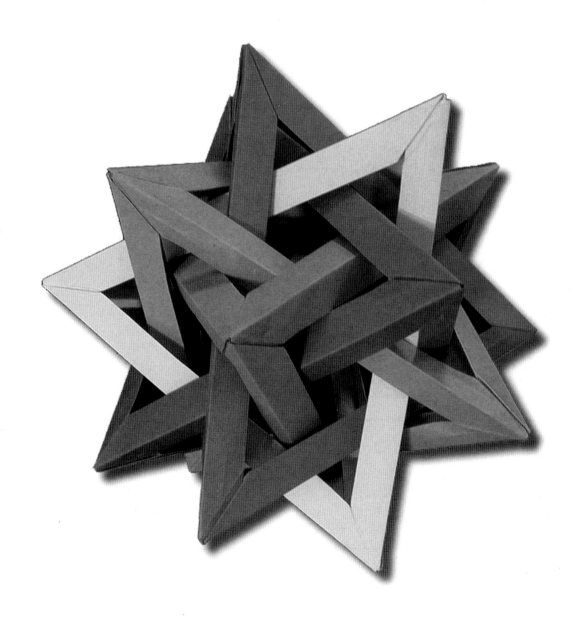

✕ Edward Mullan ✕

Contents

Introduction

Maths 1

Maths 2

Maths 3

Practice Exam

Introduction

Ptolomy I (circa 300BC) once spoke with the mathematician Euclid concerning his mathematical work, Elements, which ran to 13 books. Wanting to learn maths but without going through all the books he asked Euclid if there was not a short-cut to getting the information. Euclid, in disgust, famously replied, "There is no Royal Road to Mathematics!"

By this he meant that there was no route to achievement except through understanding and effort on the part of the person wanting to achieve.

This book contains all the facts you need to pass Higher Mathematics. It contains facts, explanations and worked examples. It suggests contexts where much of the maths is encountered in real-life. It provides structure to help you prepare for assessment. There is, however, no Royal Road to learning. Having the book is just the first step on the way to success.

Read the explanations, try the examples and then look at the working. Did you give as full an answer as is in the given solution?

Effort and diligence are keywords. Revisit the topics regularly. If you don't use it you lose it.

Hard won knowledge can get irretrievably stored away in the recesses of the brain if it is not regularly recalled. When the unit tests come round is a good time to do this. Don't just look at the minimal outcomes for the unit test. Take the opportunity to revise the topics thoroughly. This will pay dividends later when you come to the end of the course and have to revise the whole syllabus prior to the final. Remember also that at this time you will have the burden of other subjects to revise.

The 100 m Olympic gold medal winner didn't get there by checking the night before that he could still run. Treat memory as a muscle that needs regular exercising.

If you still don't understand something after you have read it and tried your best, then you should ask your teacher for help. Anything in this book could turn up in the exam. Sure as fate, the thing you don't ask about will be the thing that turns up. There is an old adage, referred to as "Murphy's law", which states, "Anything that can go wrong, will!" This law is universally true.

There are a couple of points about conventions in the book that are worth a mention:
The decimal point is mid-line. So 3·6 reads as "three point six". I know this conflicts with many calculators and spreadsheets but for reasons of clarity the form 3.6 is used to mean "3×6". This avoids confusion in the hand-written forms found in algebra and calculus where $\times$ and x are easily mistaken for each other. I hope you agree that $3.5x$ is less confusing and less ambiguous that $3 \times 5x$ which, when hand-written, could be confused with $3x5x$. It also adds to the clarity in the working when dealing with the chain rule. Graphs have been created on Excel and some compromises have been made as to the labeling. For example $y = 3\log(x/2)(\text{base } 2)$ for $y = 3\log_2\left(\frac{x}{2}\right)$ or $y = 2^\wedge x$ for $y = 2^x$ or $y = 2x^2 + 3x + 4$ for $y = 2x^2 + 3x + 4$. Such incidents have been thought preferable to no labeling where it helps rather than hinders understanding.

Finally you'll succeed in a subject if you find it interesting. Let's change that last phrase … if you make it interesting. If it still exists after thousands of years it must be both useful and interesting. You might just have to look for it a bit. There are many books on recreational maths, puzzles, paradoxes and curios. Go to the library and get some out. For some reason these books tend to be found in the Games section rather than the maths section. Look for names such as Martin Gardner, Sam Loyd, Henry Dudeney and many others. Rob Eastaway and Jeremy Wyndham are modern names to look out for. Their books "How long is a piece of string" and "Why do buses come in threes" offer an insight as to where maths can be found in everyday life.

Good luck.

Gradient

Definitions

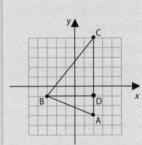

The gradient is a measure of how much a line slopes. It is defined by:

$$\text{Gradient} = \frac{\text{vertical height}}{\text{horizontal distance}}$$

This definition doesn't make a distinction between going up or down the hill.

For a horizontal line the gradient is zero.
For a vertical line the gradient is undefined.

Higher definition

A line passes through the two points $A(x_a, y_a)$ and $B(x_b, y_b)$.
Its gradient is defined by:

$$m_{AB} = \frac{y_b - y_a}{x_b - x_a}; \quad x_b \neq x_a$$

A line in the x-direction ... $y_b = y_a \Rightarrow m_{AB} = 0$
A line in the y-direction ... $x_b = x_a \Rightarrow m_{AB}$ is undefined.

Examples:

A(2, –3), B(–3, –1), C(2, 5) and D(2, –1) are four points as shown.

$$m_{AB} = \frac{-1 - (-3)}{-3 - 2} = \frac{2}{-5} = -\frac{2}{5} \quad \bigg| \quad m_{BC} = \frac{5 - (-1)}{2 - (-3)} = \frac{6}{5} \quad \bigg| \quad m_{BD} = \frac{-1 - (-1)}{2 - (-3)} = \frac{0}{5} = 0$$

$$m_{AC} = \frac{5 - (-3)}{2 - 2} = \frac{8}{0} \; \dots \text{ undefined}$$

The gradient of AB is *negative*. It slopes *downwards* in the x-direction.
The gradient of BC is *positive*. It slopes *upwards* in the x-direction.

Parallel lines

When two lines are parallel their gradients are the same.
The converse is also true, i.e. when the gradients are the same the lines are parallel.

Example
Which two lines are parallel?

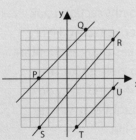

$$m_{PQ} = \frac{5 - 0}{2 - (-3)} = \frac{5}{5} = 1 \quad \bigg| \quad m_{SR} = \frac{4 - (-5)}{5 - (-3)} = \frac{9}{8} \quad \bigg| \quad m_{TU} = \frac{-1 - (-5)}{5 - 1} = \frac{4}{4} = 1$$

$$m_{TU} = m_{PQ} \Rightarrow \text{TU is parallel to PQ}$$

$$m_{SR} \neq m_{PQ} \Rightarrow \text{SR is not parallel to PQ}$$

Perpendicular lines

When two lines are perpendicular the product of their gradients is –1.
The converse is also true, i.e. if the product of their gradients is –1 then two lines are perpendicular.

Note: *Exceptions to the rule: A line in the x-direction is perpendicular to a line in the y-direction … because the gradient of a line in the y-direction is undefined we can't start to discuss the product of their gradients.*

Example

Which line is perpendicular to EF?

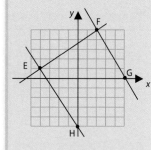

$$m_{EF} = \frac{5-1}{2-(-4)} = \frac{4}{6} = \frac{2}{3} \qquad m_{FG} = \frac{0-5}{5-2} = \frac{-5}{3} = -\frac{5}{3} \qquad m_{EH} = \frac{-5-1}{0-(-4)} = \frac{-6}{4} = -\frac{3}{2}$$

$$m_{EF}\, m_{FG} = \frac{2}{3} \times -\frac{5}{3} = -\frac{10}{9} \neq -1 \Rightarrow \text{FG is not perpendicular to EF}$$

$$m_{EF}\, m_{EH} = \frac{2}{3} \times -\frac{3}{2} = -\frac{6}{6} = -1 \Rightarrow \text{EH is perpendicular to EF}$$

Top Tip

If the gradient of a line is *a/b* then the gradient of its perpendicular is *–b/a*.

Problem solving

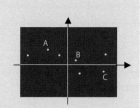

Example 1

An astronomer photographs the 'Plough' constellation in the night sky.
He believes the stars at A(–6, 4), B(2, 1) and C(10, –2) lie in a straight line.
Prove that they do.

Response

$$m_{AB} = \frac{1-4}{2-(-6)} = \frac{-3}{8} \qquad m_{BC} = \frac{-2-1}{10-2} = \frac{-3}{8} \qquad m_{AB} = m_{BC} \Rightarrow \text{AB is parallel to BC}$$

But B lies on both lines. So AB and BC must lie in the same line.
So A, B and C lie on the same line.

Example 2

P(–2, 3), Q(2, 6), R(5, 2), and S(1, –1) are the vertices of a quadrilateral with all four sides of equal length.
Prove that it is a square and not just a rhombus.

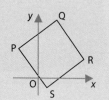

Response

$$m_{PQ} = \frac{6-3}{2-(-2)} = \frac{3}{4} \qquad m_{PS} = \frac{-1-3}{1-(-2)} = \frac{-4}{3}$$

$$m_{PQ}\, m_{PS} = \frac{3}{4} \times -\frac{4}{3} = -\frac{12}{12} = -1 \Rightarrow \text{PQ is perpendicular to PS}$$

So $\angle QPS = 90°$. So PQRS is a square.

Top Tip

To prove A, B, and C are collinear in 2D:
(i) prove $m_{AB} = m_{BC}$;
(ii) state that AB and BC have a common point, i.e. B.

Angles and Distances

Problem solving

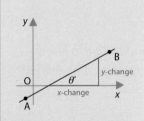

A straight line, AB, makes an angle of $\theta°$ with the positive direction of the x-axis. You can see from the diagram that this means $m_{AB} = \tan\theta°$

Examples

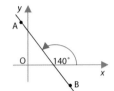

$m_{AB} = \tan 36° = 0.73$

$m_{AB} = \tan 140° = -0.84$

In a similar fashion, if you know the gradient of a line, you can work out the angle that it makes with the x-axis.

Examples

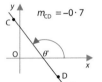

$m_{AB} = \tan\theta° = 0.6$

$\Rightarrow \theta = 31.0$

$m_{CD} = \tan\theta° = -0.7$

$\Rightarrow \theta = -35.0$ or $180 + (-35.0) = 145.0$

$\Rightarrow \theta = 145.0$ (see diagram)

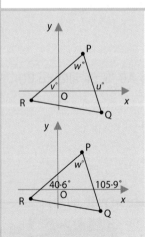

Top Tip

A sketch always helps … but guard against making it exact (e.g. no grids). No marks can be awarded for answers obtained from a scale drawing.

Problem solving

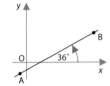

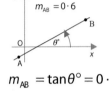

A triangle has vertices P(3, 5), Q(5, -2) and R(-4, -1). What is the size of the angle ∠QPR?

Response

A rough sketch helps you plan your steps:

$m_{PQ} = \dfrac{-2-5}{5-3} = \dfrac{-7}{2} = \tan u° \Rightarrow u° = -74.1°$ or $105.9°$

$m_{PR} = \dfrac{-1-5}{-4-3} = \dfrac{-6}{-7} = \tan v° \Rightarrow v° = 40.6°$

Simple geometry now tells us that

$w = 105.9° - 40.6 = 65.3°$

∠QPR = $65.3°$

Top Tip

The angle a line makes with the x-axis is the angle whose tangent is m.

Distances

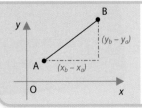

Given the points A (x_a, y_a) and B (x_b, y_b) we can calculate the distance AB using the formula:

$$AB = \sqrt{(x_b - x_a)^2 + (y_b - y_a)^2}$$

This is simply based on Pythagoras' Theorem.

Handling surds

Don't be in a rush to turn a surd into a decimal. Working with exact answers is always better.

Example 1
What is the distance between P(−10, 4) and Q(14, 11)?
Response
Using the distance formula: $PQ = \sqrt{(14-(-10))^2 + (11-4)^2} = \sqrt{24^2 + 7^2} = \sqrt{625} = 25$

Example 2
What is the distance between A(−1, 4) and B(1, 8)?
Response
Using the distance formula: $AB = \sqrt{(1-(-1))^2 + (8-4)^2} = \sqrt{20} = \sqrt{4 \times 5} = 2\sqrt{5}$

Note: *In most cases we would leave the answer as a surd, simplified if possible. If we are asked to prove two lines are equal, it will not be good enough to say they are equal when rounded off.*

Example 3
What kind of triangle has vertices P(−4, 7), Q(8, 1) and R(−3, −6)?
Response
Calculating the lengths of the sides:

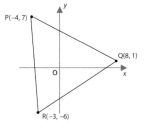

$$PQ = \sqrt{(8-(-4))^2 + (1-7)^2} = \sqrt{180}$$

$$QR = \sqrt{(-3-8)^2 + (-6-1)^2} = \sqrt{170}$$

$$RP = \sqrt{(-4-(-3))^2 + (7-(-6))^2} = \sqrt{170}$$

Two sides are the same, so triangle PQR is isosceles.

Note: *To the nearest whole number each side is 13 units. And the diagram does suggest it might be equilateral. Keep your answers in surd form to avoid coming to such mistaken conclusions.*

The midpoint

If C is the point midway between A (x_a, y_a) and B (x_b, y_b) then the coordinates of C are:

$$x_c = \frac{x_a + x_b}{2}; y_c = \frac{y_a + y_b}{2}$$

Example
Given the points A(−3, 4) and B(5, 8), find C, the midpoint of AB.
Use the gradient and distance formulae to confirm that C is indeed the midpoint.

Response
C is the point $\left(\dfrac{-3+5}{2}, \dfrac{4+8}{2}\right)$ i.e. C(1, 6) is the point.

$$AC = \sqrt{(1-(-3))^2 + (6-4)^2} = \sqrt{20}$$

$$BC = \sqrt{(1-5)^2 + (6-8)^2} = \sqrt{20}$$

AC = BC so C is equidistant from A and B.

$$m_{AC} = \frac{6-4}{1-(-3)} = \frac{2}{4} = \frac{1}{2}$$

$$m_{BC} = \frac{6-8}{1-5} = \frac{-2}{-4} = \frac{1}{2}$$

Gradients are the same and C is a common point. So A, B and C are collinear. So C is the midpoint of AB.

The Equation of a Line, part 1

The equation $y = mx + c$ represents a straight line with gradient m and y-intercept c.

The equation $y = c$ represents a horizontal line with y-intercept c.

The equation $x = d$ represents a vertical line with x-intercept d.

The general equation

Every straight line has an equation of the form $ax + by + c = 0$
(a, b, c are constants and a and b are not both zero).

Case 1: $a = 0$ $by + c = 0$ $\Rightarrow y = -\dfrac{c}{b}$ (a constant)

       e.g. $y = 4$ … a horizontal line.

Case 2: $b = 0$ $ax + c = 0$ $\Rightarrow x = -\dfrac{c}{a}$ (a constant)

       e.g. $x = 5$ … a vertical line.

Case 3: $a, b \neq 0$ $ax + by + c = 0$ $\Rightarrow y = -\dfrac{a}{b}x - \dfrac{c}{b}$

       … a straight line with gradient $-\dfrac{a}{b}$ and y-intercept $-\dfrac{c}{b}$

       e.g. $y = 2x + 1$

Top Tip

A point which satisfies $ax + by + c = 0$ is ON the line.
A point which satisfies $ax + by + c > 0$ is ABOVE the line.
A point which satisfies $ax + by + c < 0$ is BELOW the line.

The line that passes through (x_1, y_1) with gradient m

If you know one point that the line passes through, (x_1, y_1), and you know its gradient, m, then you can write down its equation: $y - y_1 = m(x - x_1)$

Example 1

What is the equation of the line that passes through the point (3, 5) with a gradient of −1?

Response

The line which passes through (3, 5) and has a gradient of −1 has the equation $y - 5 = -1(x - 3)$.

When solving problems you may have to work out a point and a gradient.

Example 2

A line passes through the points A(−1, 2) and B(5, 14).
Find the equation of the perpendicular bisector of AB.

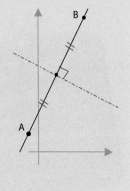

We need a point and a gradient: the midpoint of AB lies on the desired line.

This has coordinates $\left(\dfrac{-1+5}{2}, \dfrac{2+14}{2}\right)$ i.e. $(2, 8)$

It is at right angles to AB:

$$m_{AB} = \frac{14 - 2}{5 - (-1)} = \frac{12}{6} = 2 \implies m_\perp = -\frac{1}{2}$$

Thus its equation is $y - 8 = -\frac{1}{2}(x - 2)$

… which, if you wish, simplifies to $x + 2y - 18 = 0$

The line which passes through (x_1, y_1) and (x_2, y_2)

If you know two points that the line passes through, (x_1, y_1) and (x_2, y_2), you can work out its gradient, m. Then using this and the point (x_1, y_1), you can write down its equation:

$$m = \frac{y_2 - y_1}{x_2 - x_1} \quad \text{then} \quad y - y_1 = m(x - x_1)$$

Example

Find the equation of the line which passes through P(1, 4) and Q(7, −14).

Response

$$m_{PQ} = \frac{y_q - y_p}{x_q - x_p} = \frac{-14 - 4}{7 - 1} = \frac{-18}{6} = -3 \quad \text{and the equation is } y - 4 = -3(x - 1).$$

The Equation of a Line, part 2

Lines in a triangle

When problem solving in this field you should know certain lines:

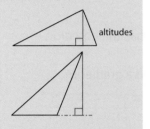
a median

a) The median

This joins the midpoint of one side of a triangle to the opposite vertex.
A triangle has three medians.
The three medians all intersect at the same point.
We say they are *concurrent*.

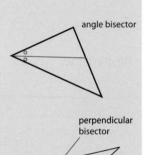
altitudes

b) The altitude

This is perpendicular to a side and passes through the opposite vertex.
You may have to extend the side.
A triangle has three altitudes.
The altitudes are *concurrent*.

angle bisector

c) Angle bisector

This passes through a vertex cutting the angle in half.
A triangle has three angle bisectors.
The angle bisectors are *concurrent*.

perpendicular bisector

d) Perpendicular bisector

This passes through the midpoint of a side
and is at right angles to the side.
A triangle has three perpendicular bisectors.
The perpendicular bisectors are *concurrent*.

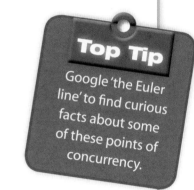
Top Tip

Google 'the Euler line' to find curious facts about some of these points of concurrency.

From each you can glean either two points or a point and a gradient … so getting the equation of each of these lines should be easy.

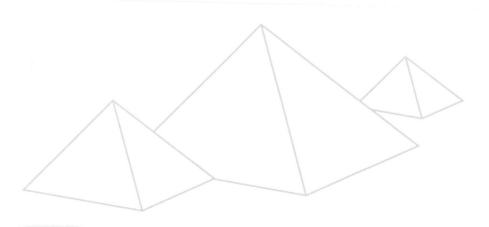

Intersecting lines

If two lines have the same gradient then they are parallel. Otherwise they intersect. We find the point of intersection by solving the equations of the lines simultaneously.

Example 1
Where do the lines $y = 2x + 5$ and $y = -x + 2$ intersect?

Response
Equate the expressions for y.
$2x + 5 = -x + 2$
$3x = -3$
$x = -1$
$y = 2x + 5 = -2 + 5 = 3$
The lines intersect at $(-1, 3)$

Example 2
Where do the lines $2x + 3y = 8$ and $3x - 5y = -7$ intersect?

Response
$2x + 3y = 8$ … ①
$3x - 5y = -7$ … ②

① × 5: $10x + 15y = 40$ … ③
② × 3: $9x - 15y = -21$ … ④
③ + ④: $19x = 19$
          $\Rightarrow x = 1$
substitute into ①: $2 + 3y = 8$
              $\Rightarrow \quad y = 2$
              The point of intersection is $(1, 2)$

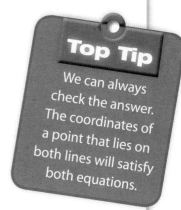

Top Tip

We can always check the answer. The coordinates of a point that lies on both lines will satisfy both equations.

Concurrency

We can prove three lines are concurrent by finding the point where two of the lines intersect, and then showing that this point satisfies the equation of the third line.

Example
Show that the lines $2x + 3y = 8$, $3x - 5y = -7$ and $y = 3x - 1$ are concurrent

Response
From the preceding example we see that the lines $2x + 3y = 8$, $3x - 5y = -7$ intersect at $(1, 2)$.
Substitute $x = 1$ into the third line, $y = 3x - 1 … y = 3.1 - 1 = 2$
So the point $(1, 2)$ lies on $y = 3x - 1$ too.
So the three lines all pass through $(1, 2)$.
They are concurrent.

Functions

Definitions

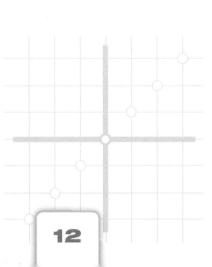

A **function**, f, is a rule which maps each member, x, of one set (called the **domain**) onto its image, $f(x)$, in another set (called the **range**).

The arrow diagram illustrates a simple mapping f, which uses the rule $x \rightarrow x - 1$ on the first five odd numbers to produce the first five even numbers.

We can let people know the rule by writing $f(x) = x - 1$

The *inverse* of a function 'undoes' the effect of the function. It is itself a function.

The inverse of the function f is denoted by f^{-1}.
If $f(x) = y$ then $f^{-1}(y) = x$ – look at your calculator to see its use.
The inverse of the sin function is $\sin^{-1}$; of the cos function is $\cos^{-1}$; of the tan function is $\tan^{-1}$.

If you know the graph of a function then you obtain the graph of the inverse by reflection in the line $y = x$.

Restrictions

There must be an image for every point in the domain, otherwise it is not a function.

The domain is often restricted so that a function can be defined.

Calculators are programmed to give out an ERROR report if you try to use a function with values not in the domain.

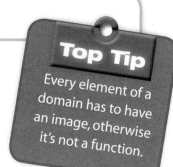

Top Tip

Every element of a domain has to have an image, otherwise it's not a function.

What does your calculator do when you ask for √(-1) or tan(90°) or the value of $1/x$ when $x = 0$?

In the exam you may be asked to suggest a suitable domain for a function. Just remember that:

(i)   you can't divide by zero so, for example, you can work out the value of $f(x) = \dfrac{3}{x - 5}$ for any value of x except 5 (where you will find yourself trying to work out $3 \div 0$).

(ii)   you can't find the square root of a negative quantity. You can work out the value of $f(x) = \sqrt{x - 4}$ so long as $x \geq 4$.

(iii)   in the range $0 \leq x \leq 360$, you can't get tan(90°) nor tan(270°), so given, say, $f(x) = \tan(x + 30)°, 0 \leq x \leq 360$ then $x \neq 60, 240$.

(iv) $f(x) = \log_a x$ requires that $x > 0$.

(v)   if you use your calculator to find the inverse sine or cosine then you will get an error message if you try to use values outside the range $-1 \leq x \leq 1$.

Composite functions

If a function g is applied to the range of a function f, the 'overall effect' is a function h called a composite function.

So if $f(x) = x + 1$ and $g(x) = x^2$ then $h(x) = g(f(x)) = g(x + 1) = (x + 1)^2 = x^2 + 2x + 1$.
Note that $f(g(x)) = f(x^2) = x^2 + 1$, which is different from $g(f(x))$.

Top Tip

In general
$f(g(x)) \neq g(f(x))$

Graphs, part 1

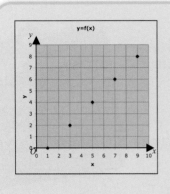

We can draw a graph of the function by plotting y against x where $y = f(x)$

You should be able to recognise some basic functions by their graphs and be able to suggest what their equation might be.

Polynomials

$y = x^2$

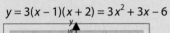

$y = 3(x - 1)(x + 2) = 3x^2 + 3x - 6$

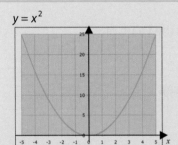

The equation of a quadratic takes the form $y = a(x - b)(x - c)$. From the graph we can pick out $b = -2$ and $c = 1$, where it cuts the x-axis.
We can also see that when $x = 0$,
$y = -6, -6 = a(0 - 1)(0 + 2) \implies a = 3$
$y = 3(x - 1)(x + 2) = 3x^2 + 3x - 6$

$y = x^3$

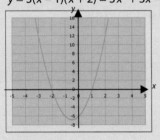

$y = 3(x - 1)(x + 2)(x + 3)$

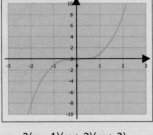

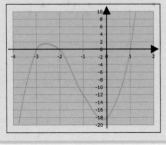

The equation of a cubic takes the form
$y = a(x - b)(x - c)(x - d)$.
From the graph we can pick out $b = -3, c = -2$ and $d = 1$ where it cuts the x-axis.
We can also see when $x = 0$,
$y = -18$ so $-18 = a(0 - 1)(0 + 2)(x + 3) \implies a = 3$.
$y = 3(x - 1)(x + 2)(x + 3) = 3x^3 + 12x^2 + 3x - 18$

Trigonometric

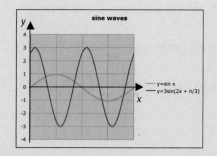

The equation takes the form $y = a \sin(bx + c)$.

The amplitude, $a = 3$, is the height of the wave above and below the x-axis.

There are 2 waves in the interval $0 \le x \le 2\pi$ giving us $b = 2$ (Divide 2π by b to get the *period* giving us a period of π. The pattern repeats every π units.)

c can be found if we are told where the curve cuts the x-axis by solving $3\sin(2x + c) = 0$ or if we are told where it cuts the y-axis by solving $3\sin c = y$

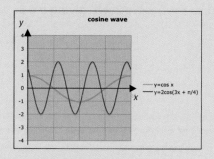

The equation takes the form $y = a \cos(bx + c)$.

The amplitude, $a = 2$, is the height of the wave above and below the x-axis.

There are 3 waves in the interval $0 \le x \le 2\pi$ giving us $b = 3$ (Divide 2π by b to get the period: $2\pi/3$. The pattern repeats every $2\pi/3$ units.)

c can be found if we are told where the curve cuts the x-axis by solving $2\cos(3x + c) = 0$ or if we are told where it cuts the y-axis by solving $2\cos c = y$

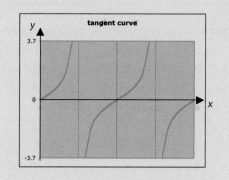

The tangent curve has a period of π radians.

Top Tip

Working in degrees with basic trig is Standard Grade, working in radians is Higher.

Graphs, part 2

Exponential curves

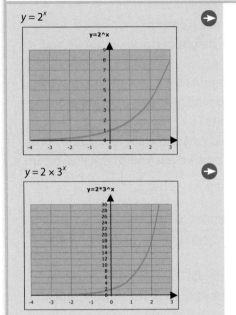

The basic exponential graph has an equation $y = a^x$ where a is a constant.
The graph passes through $(0, 1)$. All graphs with equation $y = a^x$ do that.
It also passes through $(1, 2)$ and so $2 = a^1$, giving $a = 2$.

The equation is $y = 2^x$.

This graph has an equation of the form $y = a \times b^x$.
It passes through $(0, 2)$ and so $2 = a \times b^0$, giving $a = 2$.
It also passes through $(1, 6)$ and so $6 = a \times b^1 = 2 \times b^1$, giving $b = 3$.

The equation is $y = 2 \times 3^x$.

Logarithmic curves

The basic log graph has an equation $y = \log_a x$ where a is a constant.

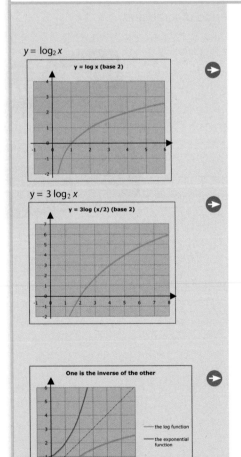

The top graph passes through $(1, 0)$, as do all graphs with equation $y = \log_a x$. It also passes through $(2, 1)$ and so $1 = \log_a 2$, giving $a = 2$ since it is a rule that $\log_a a = 1$. The equation is $y = \log_2 x$.

Suppose we're told that the middle graph has an equation of the form $y = b \times \log_2(cx)$. It passes through $(2, 0)$ and so:
$0 = b \times \log_2(2c) \implies \log_2(2c) = 0$. And since $\log_a 1 = 0 \implies c = {}^1/_2$

It also passes through $(4, 3)$ and so:
$3 = b \times \log_2(4 \times {}^1/_2) \implies b \log_2 2 = 3$
$\implies b = 3$
The equation is $y = 3 \log_2({}^x/_2)$

The graph of the exponential function is the reflection of the log function in the line $y = x$.

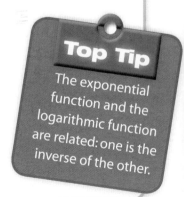

Top Tip

The exponential function and the logarithmic function are related: one is the inverse of the other.

Related functions

If we have the graph of $y = f(x)$ then:

(i) to draw the graph $y = -f(x)$ we need only reflect in the x-axis:

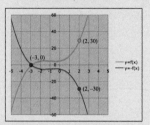

(ii) to draw the graph $y = f(-x)$ we need only reflect in the y-axis:

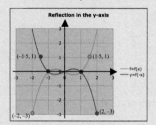

You should also know the following related functions:

(iii) $y = f(x) + a$

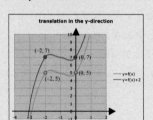

(iv) $y = f(x - a)$

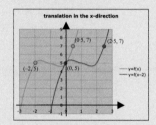

(v) $y = af(x)$

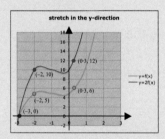

(vi) $y = f(ax)$

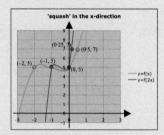

In each illustration $a = 2$. Of course the '2' could be any number.
You could be asked to combine these transformations.

(vii) You can graph the inverse, $y = f^{-1}(x)$, by reflecting $y = f(x)$ in the line $y = x$.
Here are three functions and their inverses:

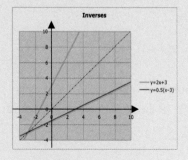

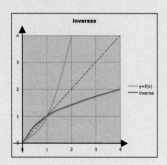

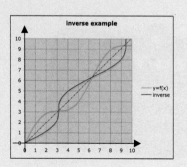

Graphs, part 3

Maxima and minima of functions

(i) Particular cases

Certain functions can't get any smaller than a particular value (its minimum) or any larger than a particular value (its maximum). For example:

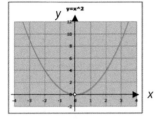
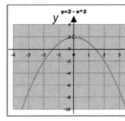
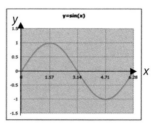
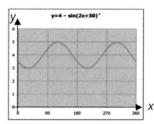

$y = x^2$ can't get smaller than 0

$y = 2 - x^2$ can't get larger than 2
… because x^2 can't get smaller than 0.

$y = \sin x$ can't get larger than 1 or smaller than −1
i.e. $-1 \le \sin x \le 1$

$y = 4 - \sin(2x + 30)°$ can't get larger than $4 - (-1)$ or smaller than $4 - 1$
i.e. $3 \le 4 - \sin(2x + 30)° \le 5$
… because
$-1 \le \sin(2x + 30)° \le 1$

(ii) Completing the square

We know $(x + a)^2 = x^2 + 2ax + a^2$

Rearranging, we get $x^2 + 2ax = (x + a)^2 - a^2$

This rearrangement is known as 'completing the square'.

This is handy whenever we want to simplify an expression containing an x^2 term and an x term.

Example 1

Simplify $x^2 + 6x$.

We note that the coefficient of x, $2a$ is 6 and mentally work out that $a = 3$.

We can then write down:

$x^2 + 6x = (x + 3)^2 - 3^2 = (x + 3)^2 - 9$.

Example 2

Simplify $x^2 - 8x - 3$.

Note mentally that $a = -4$.

$x^2 - 8x - 3 = (x - 4)^2 - (-4)^2 - 3 = (x - 4)^2 - 19$.

Example 3

Simplify $2x^2 - 8x - 3$.

The 'trick' works when x^2 has a coefficient of 1.

Take 2 out as a common factor from the terms containing x:

$2[x^2 - 4x] - 3$

Complete the square inside the brackets:

$2[(x - 2)^2 - (-2)^2] - 3 = 2[(x - 2)^2 - 4] - 3$

Now get rid of the brackets again:

$= 2(x - 2)^2 - 8 - 3$

$= 2(x - 2)^2 - 11$

An exam question will more than likely be expressed as follows:

Example 4

Express $3x^2 + 6x + 5$ in the form $a(x + b)^2 + c$.

Response

$3x^2 + 6x + 5 = 3(x^2 + 2x) + 5$

$= 3\left((x + 1)^2 - 1^2\right) + 5$

$= 3(x + 1)^2 - 3 + 5 = 3(x + 1)^2 + 2$

Why is this manipulation useful?
Example 5 on page 19 illustrates one of the main reasons

Top Tip

Differentiation can be used to find maxima and minima in Higher but not in Example 5.

Example 5

a) Express $x^2 + 6x + 11$ in the form $(x + a)^2 + b$.

b) Hence or otherwise find the largest value that

the expression $\dfrac{6}{x^2 + 6x + 11}$

can take, and the corresponding value of x.

Response

a) $x^2 + 6x + 11 = (x + 3)^2 - 3^2 + 11 = (x + 3)^2 + 2$

b) $\dfrac{6}{x^2 + 6x + 11} = \dfrac{6}{(x + 3)^2 + 2}$

We know that the smallest a perfect square can be is zero.

So the smallest $(x + 3)^2$ can equal is 0. This will happen when $x = -3$.

So the smallest $(x + 3)^2 + 2$ can equal is 2.

The expression $\dfrac{6}{(x + 3)^2 + 2}$ will be at its

largest when the denominator is at its smallest

i.e. $\dfrac{6}{2} = 3$

The largest value the expression can achieve is 3 when $x = -3$.

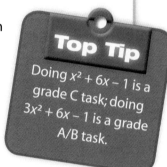

Top Tip

Doing $x^2 + 6x - 1$ is a grade C task; doing $3x^2 + 6x - 1$ is a grade A/B task.

(iii) Trigonometric functions

See the chapter on the wave function (pages 72-75) to discover why the maximum and minimum of a function of the form $f(x) = a \sin x + b \cos x$ are $\sqrt{a^2 + b^2}$ and $-\sqrt{a^2 + b^2}$ respectively.

Problem solving

Whenever a question appears in context in this field it will generally involve maxima and minima.

Example

Relative to a particular set of axes, the arch of the Tyne Bridge is a

parabola with equation $h = -\dfrac{1}{8}\left(x^2 - 9x + 8\right)$ where h is the

height of the arch measured in suitable units.

What is the highest point and for what value of x is it achieved?

Response

Completing the square we get

$h = -\dfrac{1}{8}\left(x^2 - 9x + 8\right) = -\dfrac{1}{8}\left((x - 4\cdot5)^2 - 4\cdot5^2 + 8\right)$

$= -\dfrac{1}{8}\left((x - 4\cdot5)^2 - 12\cdot25\right)$

The smallest a perfect square can be is zero, so:

$h_{max} = 12\cdot25 \div 8 = 1\cdot53$ units (to 2 d.p.), achieved when $x = 4\cdot5$.

The Derived Function, part 1

The gradient at a point

The gradient of the line passing through $A(x_a, y_a)$ and $B(x_b, y_b)$ is $m_{AB} = \dfrac{y_b - y_a}{x_b - x_a}$

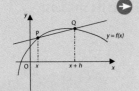

P is a point on the curve $y = f(x)$. Its coordinates are $(x, f(x))$.
Q is a point further along the curve with x-coordinate $x + h$ (where h is how much more to the right Q is than P). Its coordinates are $(x + h, f(x + h))$.

$$m_{PQ} = \frac{f(x + h) - f(x)}{x + h - x} = \frac{f(x + h) - f(x)}{h}$$

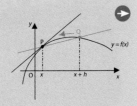

Allow h to get smaller and smaller: Q will head towards P.
PQ will get more and more like the tangent at P;
the gradient of PQ should become the gradient of the tangent at P.

We define the gradient of the curve at the point P as the gradient of the tangent at the point and denote it by $f'(x)$ where

$$f'(x) = \lim_{h \to 0} \frac{f(x + h) - f(x)}{h}$$

We could use this to *estimate* the gradient of the curve at a particular point by taking h to be a small quantity, say $h = 0.0001$.

Example
Estimate the gradient of the curve $y = 3x^2 - 8x$ at the point where $x = 2$.

Response

$$f'(2) \approx \frac{\left[3(2 + 0.0001)^2 - 8(2 + 0.0001)\right] - \left[3.2^2 - 8.2\right]}{0.0001} = \frac{0.00040003}{0.0001} = 4.0003$$

i.e. if $f(x) = 3x^2 - 8x \Rightarrow f'(2) \approx 4$

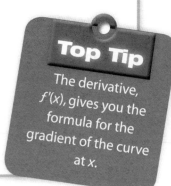

Top Tip

The derivative, $f'(x)$, gives you the formula for the gradient of the curve at x.

The gradient at a point ... an exact method

Top Tip

In words, this means 'Multiply by the power then reduce the power by 1.'

Isaac Newton devised an exact method for finding a *formula* for $f(x)$:
$$f(x) = ax^n \Rightarrow f'(x) = nax^{n-1}$$

This method is called **differentiation**:
$f'(x)$ is called the *derivative* or *derived function*;
what you get is a formula for the gradient of the curve at x.

Example 1
What is the gradient of the curve $y = 3x^2 - 8x + 3$ at the point where $x = 2$?

Response

$f(x) = 3x^2 - 8x + 3 \Rightarrow f'(x) = 6x - 8$ We differentiate each term using the rule.
$\Rightarrow f'(2) = 6.2 - 8 = 4$

Compare this with the approximate value on page 20.

Just as we write $y = f(x)$ to denote the graph of a function, we use another notation, devised by a mathematician called Leibniz, to denote the derived function, so:

$$\frac{dy}{dx} = f'(x) \dots \quad \frac{dy}{dx} \quad \text{is read as 'dy by dx' or}$$
'the change in y with respect to x'.

Example 2
Find the change in y with respect to x when $y = \dfrac{x^3 + 1}{x}$

Response
Firstly, split it up: $y = \dfrac{x^3}{x} + \dfrac{1}{x} = x^2 + x^{-1}$

Now, use the rule: $\dfrac{dy}{dx} = 2x + (-1)x^{-2}$

We might be asked to find the point when the gradient is a given value.

Example 3
At what points does the tangent to the curve $y = 2x^3 - 3x^2 + 6x + 1$ have a gradient of 18?

Response

$\dfrac{dy}{dx} = 6x^2 - 6x + 6 \qquad \dfrac{dy}{dx} = 18 \Rightarrow 6x^2 - 6x + 6 = 18$

$\Rightarrow x^2 - x - 2 = 0 \Rightarrow (x + 1)(x - 2) = 0$

$\Rightarrow x = -1$ or $2 \Rightarrow y = -10$ or 17

The tangent has a gradient of 18 at $(-1, -10)$ and $(2, 17)$.

The Derived Function, part 2

The behaviour of a function

We can tell a lot by looking at the gradient of a function:

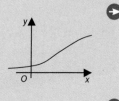

If $f'(x) > 0$ … the gradient is positive
… the function is increasing
See how the curve climbs as it moves to the right.

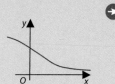

If $f'(x) < 0$ … the gradient is negative
… the function is decreasing.
See how the curve falls as it moves to the right.

If $f'(x) = 0$ … the gradient is zero
… the function is neither increasing nor decreasing
… we say it is stationary.

Look at these cases. In each case, at the point A, $f'(x) = 0$

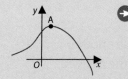

	Before A	At A	After A
$f'(x)$	+	0	−
$f(x)$	increasing	stationary	decreasing

… a maximum turning point

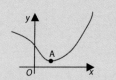

	Before A	At A	After A
$f'(x)$	−	0	+
$f(x)$	decreasing	stationary	increasing

… a minimum turning point

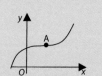

	Before A	At A	After A
$f'(x)$	+	0	+
$f(x)$	increasing	stationary	increasing

… a horizontal point of inflexion

	Before A	At A	After A
$f'(x)$	−	0	−
$f(x)$	decreasing	stationary	decreasing

… a horizontal point of inflexion

Example

Find the stationary points of the curve $y = 3x - 8x^3 + 6x^2 + 1$
and determine their nature.

Response

$\dfrac{dy}{dx} = 12x^3 - 24x^2 + 12x$

$\dfrac{dy}{dx} = 0$ at stationary points

$\Rightarrow 12x(x^2 - 2x + 1) = 0$
$\Rightarrow 12x(x - 1)^2 = 0$
$\Rightarrow x = 0$ or $x = 1$ twice.

Top Tip

$f'(x) > 0$ … function is increasing at x

$f'(x) < 0$ … function is decreasing at x

$f'(x) = 0$ … function is stationary at x

x	$\longrightarrow$	0	$\longrightarrow$	1	$\longrightarrow$
x	–	0	+	+	+
$(x-1)^2$	+	+	+	0	+
dy/dx	–	0	+	0	+
slope	\	—	/	—	/

When $x = 0$, $y = 1$ and we have a minimum turning point at $(0, 1)$.
When $x = 1$, $y = 2$ and we have an horizontal point of inflexion at $(1, 2)$.

Sketching a curve

When sketching a curve of a given equation
(i)   Find where it cuts the x-axis: $y = 0$.
(ii)   Find where it cuts the y-axis: $x = 0$.
(iii)   Find the stationary points and determine their nature: $\dfrac{dy}{dx} = 0$.

(iv)   Consider the behaviour of y for large x (positive and negative).

Example
Sketch the curve with equation $y = x(x + 9)^2$.

Top Tip

A table of signs will help you figure out the behaviour of a function.

Put this information on a sketch and 'join the dots':

Response
(i)   When $y = 0$, $x(x + 9)^2 = 0 \Rightarrow x = 0$ or $x = -9$ twice.
        The curve cuts the x-axis at $(0, 0)$ and $(-9, 0)$.
(ii)   When $x = 0, y = 0$. The curve cuts the y-axis at $(0, 0)$.
(iii) $y = x(x + 9)^2 = x^3 + 18x^2 + 81x$

$$\Rightarrow \frac{dy}{dx} = 3x^2 + 36x + 81$$

$$\frac{dy}{dx} = 0 \text{ at SPs}$$

$3x^2 + 36x + 81 = 0$
$\Rightarrow x^2 + 12x + 27 = 0$
$\Rightarrow (x + 3)(x + 9) = 0$
$\Rightarrow x = -3$ or $x = -9$

x	$\longrightarrow$	-9	$\longrightarrow$	-3	$\longrightarrow$
$x + 3$	–	–	–	0	+
$x + 9$	–	0	+	+	+
dy/dx	+	0	–	0	+
slope	/	—	\	—	/

The curve has a maximum TP at $(-9, 0)$ and a minimum at $(-3, -108)$.

(iv) $y = x(x + 9)^2$ so when x is large and positive then, since $(x + 9)^2$ will be large and positive, their product, y, will be large and positive …
        when x is large and negative then, since $(x + 9)^2$ will be large and positive, their product, y, will be large and negative.

The Derived Function, part 3

The graph of the derived function

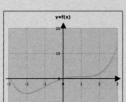

In the exam you may be asked to sketch the graph of $f'(x)$ when given the graph of $f(x)$.

Example

The graph $y = f(x)$ is as shown. It has a minimum turning point at $x = -2$ and a horizontal point of inflexion at $x = 1$.

Make a sketch of the graph $y = f'(x)$.

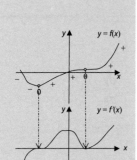

Step 1	Sketch $y = f(x)$, marking on it '+' or '–' or '0' according to whether it is increasing, decreasing or stationary
Step 2	Start a new graph below your sketch.
Step 3	Transfer the stationary points as zeros onto the new graph
Step 4	Connect these zeros by smooth curves … below the x-axis where a '–' is shown; … above the axis where a '+' is shown.

This sketch will be a reasonable indication of the behaviour of $f'(x)$.

Rates of change

As two related variables change, it is often useful to know how one is changing compared to the other. This is known as a rate.

Some examples

Mileage: miles per litre Wages: pounds per hour
Speed: kilometres per hour General: y per x

When one variable, y, can be expressed as a function of the other, $f(x)$, we can get a formula for the rate of change by differentiating.

… $\dfrac{dy}{dx}$ is the rate of change of y with respect to x (y per x).

Example

The formula: $s = 50t - 5t^2$ can be used to model the flight of a golf ball:
… t is the time in seconds since the ball was struck;
… s is the height of the ball.

Suppose we wish $v(t)$ the velocity of the ball after t seconds (i.e. the rate of change of height per second). Differentiate height with respect to time.

$$v = \frac{ds}{dt} = 50 - 10t$$

Top Tip

s stands for displacement
v stands for velocity
a stands for acceleration
t stands for time
$v = ds/dt$
$a = dv/dt$

Note that when $t = 5$, the velocity is zero. The ball has reached the top of its flight and is about to come down.

The acceleration can also be found by differentiating the velocity: $a = \dfrac{dv}{dt} = -10$

Optimisation

When a function lies in a closed interval of x, its maximum and minimum values will be at an endpoint of the interval or a turning point. Be sure to check both possibilities.

Example

A manufacturing company makes boxes from cardboard squares of side 120 cm. Squares are cut out of each corner of the square and the resulting rectangles are folded up. The small squares are of side x cm where $15 \le x \le 30$.

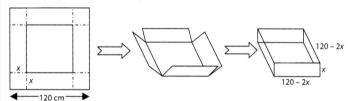

The volume of the finished article is a function of x, that is: $V(x) = x(120 - 2x)^2$. Find maximum and minimum volume possible.

Response

(i) Find the stationary points:

$$V = x(120 - 2x)^2 \Rightarrow V = 14400x - 480x^2 + 4x^3$$

$$\Rightarrow \frac{dV}{dx} = 14400 - 960x + 12x^2$$

At SPs $\dfrac{dV}{dx} = 0$ $\Rightarrow 1200 - 80x + x^2 = 0$

$$\Rightarrow (x - 20)(x - 60) = 0$$

$$\Rightarrow x = 20 \text{ or } 60$$

... $V(20) = 128\,000$ cm^3 and $V(60)$ is undefined, since $15 \le x \le 30$.

(ii) Consider the endpoints:

$V(15) = 121\,500$ cm³

$V(30) = 108\,000$ cm³

Note that the greatest volume, $128\,000$ cm³ occurs when $x = 20$ and the least volume is $108\,000$ cm³ when $x = 30$.

Note: *We don't have to do a nature table when exploring the maxima and minima in a closed interval. We know they must occur at stationary or end points.*

Sequences, part 1

Notation

Top Tip

$u_{n+1} = mu_n + c$
is a linear recurrence
relation.

Top Tip

A limit exists if
$-1 < m < 1$.
To find the limit, L,
let u_n and u_{n+1} both
equal L and solve the
equation $L = mL + c$.

A **sequence** is an ordered list of numbers.
By *ordered* we mean that the 1st, 2nd, 3rd etc terms can be identified.
The *first term* is usually referred to as u_1
The *general term*, often called the nth term, is referred to as u_n.
A sequence can be defined by expressing u_n as a function of n.

Example
List the first four terms of the sequence whose n^{th} term is $u_n = 3n - 1$.
Response
$u_1 = 3.1 - 1 = 2$; $u_2 = 3.2 - 1 = 5$; $u_3 = 3.3 - 1 = 8$; $u_4 = 3.4 - 1 = 11$
The first four terms are: 2, 5, 8, 11.
A sequence can also be defined by expressing u_{n+1} as a function of u_n and giving
the value of one term, usually u_1. However, often the term u_0 is given. It is not in the
sequence but can be used to generate u_1.
A **recurrence relation** is formed when u_{n+1} is expressed as a function of u_n and an
initial value is given.

Example
List the first four terms of the sequence defined by the recurrence relation:
$u_{n+1} = 3u_n - 1$, $u_1 = 2$.
Response
$u_1 = 2$; $u_2 = 3u_1 - 1 = 3.2 - 1 = 5$; $u_3 = 3u_2 - 1 = 3.5 - 1 = 14$;
$u_4 = 3u_3 - 1 = 3.14 - 1 = 41$
The first four terms are: 2, 5, 14, 41.

Linear recurrence relations

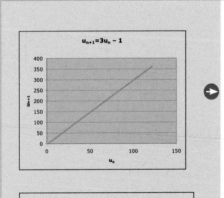

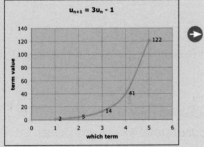

When the recurrence relation takes the form $u_{n+1} = mu_n + c$, $u_1 = a$,
we call the relation a linear recurrence relation.
If u_{n+1} is graphed against u_n a straight line is produced.

In this illustration $m = 3$ and $c = -1$.

If, on the other hand, we plot the value of the term against its
position in the sequence (1st, 2nd, etc) we get what looks like an
exponential graph.

Note that only the labelled points exist: the curve is only drawn to
highlight the relationship.

Limits

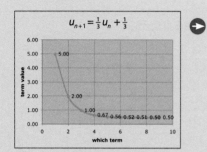

$$u_{n+1} = \tfrac{1}{3}u_n + \tfrac{1}{3}$$

A funny thing happens if the multiplier is a proper fraction, i.e. $-1 < m < 1$:

Here is the recurrence relation $u_{n+1} = \tfrac{1}{3}u_n + \tfrac{1}{3}$, $u_1 = 5$.
See how it flattens out as it progresses.
By the 10th term it looks like it is sticking at 0·5.
Check that $\tfrac{1}{3}0\cdot5 + \tfrac{1}{3} = 0\cdot5$.

When $-1 < m < 1$, the relation $u_{n+1} = mu_n + c$ heads towards a limit L.

The easiest way to find the limit is to let both u_{n+1} and u_n equal L and solve the resulting equation, $L = mL + c$.

Example
A tank at a desert oasis is initially filled with 60 litres of water.
Over the day, 10% of the contents are lost due to evaporation.
Just before noon, a valve opens and allows 5 litres to flow into the tank.
The amount of water at noon on the nth day can be modelled by
$u_{n+1} = 0\cdot9u_n + 5$, $u_1 = 60$ (If 10% is lost 90% remains.)
What do you expect to happen over time?

Response
This is a linear relationship with $m = 0\cdot9$.
$-1 < m < 1$ so a limit, L, exists.
$L = 0\cdot9L + 5$
$\Rightarrow 0\cdot1L = 5$
$\Rightarrow L = 50$

One would expect that over time the tank's contents at noon will stablise at a volume of 50 litres.

Sequences, part 2

Finding the parameters

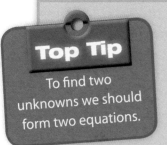

Top Tip

To find two unknowns we should form two equations.

Example
The first three terms of a linear recurrence relation are $5, 15, 35$.
Find the relationship.

Response
The relation is of the form $u_{n+1} = mu_n + c$.

Using the first two terms:	$15 = 5m + c$	... ①
Using the 2nd and 3rd term:	$35 = 15m + c$	... ②
Subtracting ① from ② :	$20 = 10m$	
	$\Rightarrow m = 2$	
Substitute in ① :	$15 = 5.2 + c$	
	$\Rightarrow c = 5$	
The recurrence relation is:	$u_{n+1} = 2u_n + 5, u_1 = 5$	

Contexts

Top Tip

These models are only defined for n **W**.

The linear recurrence relation is often used to model growth and decay. You should realise that the model only takes 'snapshots' at regular intervals in the story, at $n = 1, 2, 3$ etc. Be careful not to read too much into values between these.

Interest in the bank (growth)
Example
A saver puts £1000 into the bank. The bank offer 6% p.a. if the money is left for 5 years and a regular £100 is deposited just before the end of each year.
(i)   Form a recurrence relation to model the situation.
(ii)   How much is in the account at the end of the 3rd year?

Response
(i) $u_{n+1} = 1{\cdot}06u_n + 100, u_0 = 1000$
(ii) $u_1 = 1{\cdot}06.1000 + 100 = 1160$
         $u_2 = 1{\cdot}06.1160 + 100 = 1329{\cdot}60$
         $u_3 = 1{\cdot}06.1329{\cdot}60 + 100 = 1509{\cdot}38$
         At the end of the 3rd year there will be £1509·38.

Cooling coffee (decay)

Example

At first the contents of a coffee-maker are 80°C above room temperature.
As the coffee cools, this difference drops by 10% each minute.
How long will it be before the contents drop to 20°C above room temperature?

Response

The recurrence relation which models this situation is: $u_{n+1} = 0.9u_n$, $u_0 = 80$.

(i)   Type '80 =' into your calculator. Now the calculator thinks the last answer is 80.

(ii)   Type '0·9 × ANS ='. The calculator will return 72 … one minute has passed

(iii)   Type '='. The calculator will return 64·8 … two minutes have passed.

(iv)   Continue typing '=', keeping a note of how often you have done so.
The calculator will return 20·3349266 on the 13th press and 18·301434 on the 14th press.
The coffee will get to 20°C above room temperature during the 14th minute.

Example

Greenhouse effect (decay)

A greenhouse is initially heated to 10°C above the outside temperature.
It loses 25% of its heat through the glass each hour. At the end of each hour a heater boosts its temperature by 2°C.

(i)   Form a recurrence relation to model the situation.

(ii)   What do you expect to happen over time?

Response

(i) $u_{n+1} = 0.75u_n + 2$, $u_0 = 10$

(ii) $-1 < 0.75 < 1$ so a limit, L, exists.

$L = 0.75L + 2$

$\Rightarrow 0.25L = 2$

$\Rightarrow L = 8$

Over time the temperature inside the greenhouse will settle down to being 8°C above the outside temperature.

Top Tip

The coefficient of u_n will give you a hint as to whether it is growth (>1) or decay (<1).

Maths 1 Assessment

This completes Maths 1.
For the unit test, you will be expected to show that you can perform the following under exam conditions.
The SQA have published this list in their conditions and arrangements.

OUTCOME 1

Use the properties of the straight line.

Performance criteria
a) Determine the equation of a straight line given two points on the line or one
point and the gradient.
b) Find the gradient of a straight line using $m = \tan$.
c) Find the equation of a line parallel to and a line perpendicular to a given line.

OUTCOME 2

Associate functions and graphs.

Performance criteria
a) Sketch and identify related graphs and functions.
b) Identify exponential and logarithmic graphs.
c) Find composite functions of the form $f(g(x))$, given $f(x)$ and $g(x)$.

OUTCOME 3

Use basic differentiation.

Performance criteria
a) Differentiate a function reducible to a sum of powers of x.
b) Determine the gradient of a tangent to a curve by differentiation.
c) Determine the coordinates of the stationary points on a curve and justify their
nature using differentiation.

OUTCOME 4

Design and interpret mathematical models of situations involving recurrence
relations.

Performance criteria
a) Define and interpret a recurrence relation in the form
$u_{n+1} = mu_n + c$ (m, c constants) in a mathematical model.
b) Find and interpret the limit of the sequence generated by a recurrence
relation in a mathematical model (where the limit exists).

Top Tip

Practice tests suitable for
preparing for the unit test can
be downloaded from
www.leckieandleckie.co.uk.

Quadratic Theory, part 1

The roots of a quadratic equation

The roots of the equation $ax^2 + bx + c = 0, a \neq 0,$

are $x = \dfrac{-b \pm \sqrt{b^2 - 4ac}}{2a}$.

Top Tip

Remember: We can't divide by zero; we can't get the square root of a negative number.

Example

Solve the equation $3x^2 + 5x - 2 = 0$.

Response

By inspection $a = 3, b = 5$ and $c = -2$.

Substituting into the formula gives

$$x = \frac{-5 \pm \sqrt{5^2 - 4.3.(-2)}}{2.3} = \frac{-5 \pm \sqrt{49}}{6} = \frac{-5 \pm 7}{6}$$

So $x = {}^1/_3$ or $x = -2$.

The nature of the roots

In the quadratic formula, the term which lies under the square root sign, $b^2 - 4ac$, is called the *discriminant*. By examining it we can distinguish between the various things that can happen.

(i) $b^2 - 4ac < 0$
     We can't work out the square root: there are no real roots.

(ii) $b^2 - 4ac = 0 \dots \; x = \dfrac{-b \pm 0}{2a} = \dfrac{-b}{2a}$

      Both roots are the same (coincident).

(iii) $b^2 - 4ac > 0$ and a perfect square
      Roots are different (distinct) and rational.

(iv) $b^2 - 4ac > 0$ but not a perfect square
      Roots are distinct and irrational.

Examples

By considering the discriminant in each case determine the nature of the roots of the equations:

(i) $x^2 + x + 2 = 0$
(ii) $x^2 + 2x + 1 = 0$
(iii) $4x^2 + 13x + 9 = 0$
(iv) $x^2 + x - 1 = 0$

Top Tip

In the exam, quote the actual conditions you need – not just the fact that you are using the discriminant.

Response

(i) $a = 1, b = 1, c = 2$
     $\Rightarrow b^2 - 4ac = 1^2 - 4.1.2 = -7$
     $\Rightarrow b^2 - 4ac < 0 \;\; \Rightarrow$ no real roots

(ii) $a = 1, b = 2, c = 1$
     $\Rightarrow b^2 - 4ac = 2^2 - 4.1.1 = 0$
     $\Rightarrow b^2 - 4ac = 0 \;\; \Rightarrow$ coincident roots
     $[\dots x = -1$ (twice)$]$

(iii) $a = 4, b = 13, c = 9$
     $\Rightarrow b^2 - 4ac = 13^2 - 4.4.9 = 25$
     $\Rightarrow b^2 - 4ac > 0$ and a perfect square
     $\Rightarrow$ distinct rational roots
     $[\dots x = -1$ or $x = -{}^9/_4]$

(iv) $a = 1, b = 1, c = -1$
     $\Rightarrow b^2 - 4ac = 1^2 - 4.1.(-1) = 5$
     $\Rightarrow b^2 - 4ac > 0$ but not a perfect square
     $\Rightarrow$ distinct irrational roots

$$\left[\dots x = \frac{-1 + \sqrt{5}}{2} \text{ or } x = \frac{-1 - \sqrt{5}}{2} \right]$$

Quadratic Theory, part 2

Conditions for tangency

The line $y = px + q$ will intersect the parabola $y = ax^2 + bx + c$ when $px + q = ax^2 + bx + c$. This will simplify to a quadratic equation. Using the discriminant we can identify 3 cases:

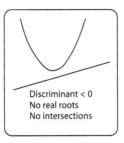

Discriminant < 0
No real roots
No intersections

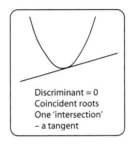

Discriminant = 0
Coincident roots
One 'intersection'
– a tangent

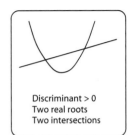

Discriminant > 0
Two real roots
Two intersections

Example 1

For what value of k is the line $y = 3x + k$ a tangent to the parabola $y = x^2 + 5x + 2$?

Response

The line and parabola intersect when $3x + k = x^2 + 5x + 2$
$\Rightarrow x^2 + 2x + 2 - k = 0$
By inspection we see that $a = 1, b = 2$ and $c = 2 - k$
The discriminant, $b^2 - 4ac = 2^2 - 4.1.(2 - k) = 4k - 4$
For coincident roots $b^2 - 4ac = 0 \Rightarrow 4k - 4 = 0$
$\Rightarrow k = 1$

Example 2

For what values of k is the line $y = 5x + 2$ a tangent to the parabola $y = x^2 + kx + 3$?

Response

The line and parabola intersect when $5x + 2 = x^2 + kx + 3$
$\Rightarrow x^2 + (k - 5)x + 1 = 0$
By inspection we see that $a = 1, b = k - 5$ and $c = 1$
The discriminant, $b^2 - 4ac$
$= (k - 5)^2 - 4.1.1$
$= k^2 - 10k + 25 - 4$
$= k^2 - 10k + 21$
For coincident roots $b^2 - 4ac = 0$
$\Rightarrow k^2 - 10k + 21 = 0$
$\Rightarrow (k - 3)(k - 7) = 0$
$\Rightarrow k = 3$ or $k = 7$

The line $y = 5x + 2$ is a tangent to both the parabolae $y = x^2 + 3x + 3$ and $y = x^2 + 7x + 3$

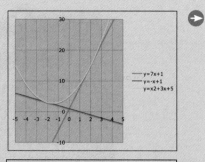

The line $y = 7x + 1$ and $y = -x + 1$ are tangents to the parabola $y = x^2 + 3x + 5$

Example 3

For what values of k is the line $y = kx + 1$ a tangent to the parabola $y = x^2 + 3x + 5$?

Response

The line and parabola intersect when $kx + 1 = x^2 + 3x + 5$

$\Rightarrow x^2 + (3 - k)x + 4 = 0$

By inspection we see that $a = 1$, $b = 3 - k$ and $c = 4$

The discriminant, $b^2 - 4ac$

$= (3 - k)^2 - 4.1.4$

$= k^2 - 6k + 9 - 16$

$= k^2 - 6k - 7$

For coincident roots $b^2 - 4ac = 0$

$\Rightarrow k^2 - 6k - 7 = 0$

$\Rightarrow (k - 7)(k + 1) = 0$

$\Rightarrow k = 7$ or $k = -1$

Intersections

On the Squinty Bridge at Finnieston in Glasgow, the arch and road are both parabolae. Using suitable axes and units, the arch has equation $y = -12x^2 + 144x - 134$ and the road has equation $y = -x^2 + 12x - 13$

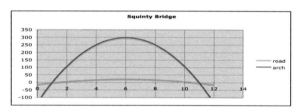

Calculate the two points where the arch intersects the road.

Response

At the intersections

$-12x^2 + 144x - 134 = -x^2 + 12x - 13$

$11x^2 - 132x + 121 = 0$

$\Rightarrow x^2 - 12x + 11 = 0$

$\Rightarrow (x - 11)(x - 1) = 0$

$\Rightarrow x = 11$ or 1

$\Rightarrow y = -2$ or -2

Points are $(11, -2)$ and $(1, -2)$

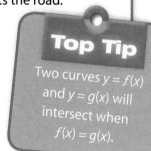

Top Tip

Two curves $y = f(x)$ and $y = g(x)$ will intersect when $f(x) = g(x)$.

Quadratic Theory, part 3

Inequalities

Example 1

Solve the inequation $2x^2 - x - 1 > 0$.

Response

Step 1 Solve the associated equation:
$$2x^2 - x - 1 = 0$$
$$\Rightarrow (2x + 1)(x - 1) = 0$$
$$\Rightarrow x = -\tfrac{1}{2} \text{ or } x = 1$$

Step 2 Make a sketch:
Because the coefficient of x^2 is positive the curve is a 'valley' rather than a 'hill'.

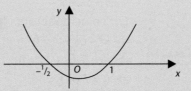

Step 3 Highlight where the function is above the axis ($f(x) > 0$)

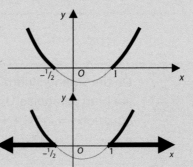

Step 4 Identify the x-values which correspond to this region

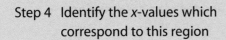

Solution: $x < -\tfrac{1}{2}$ or $x > 1$

Example 2

Solve the inequation $x^2 - 3x - 28 \leq 0$.

Response

Solving the equation:
$(x + 4)(x - 7) = 0$
$\Rightarrow x = -4$ or 7
Sketch:

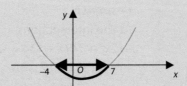

Solution: $-4 \leq x \leq 7$

Finding a quadratic with given roots

If you want to form an equation with roots $x = a$ and $x = b$, you can imagine that you are 'unsolving' the equation:

$x = a$ or $x = b$

$\Rightarrow (x - a)(x - b) = 0$

… then remove the brackets.

This will produce the simplest quadratic equation with these roots. We can find others by multiplying throughout by constants.

Any equation of the form $k(x - a)(x - b) = 0$, where k is a constant, has roots $x = a$ or $x = b$.

Example 1

Form the simplest quadratic equation whose roots are $x = 2$ and $x = -7$.

Response

$x = 2$ or $x = -7$

$\Rightarrow (x - 2)(x + 7) = 0$

$\Rightarrow x^2 + 5x - 14 = 0$

Example 2

Form three quadratic equations whose roots are $x = 2$ and $x = -7$ and sketch them on the same diagram.

Response

$x = 2$ or $x = -7$

$\Rightarrow (x - 2)(x + 7) = 0$

$\Rightarrow k(x - 2)(x + 7) = 0$

$\Rightarrow k(x^2 + 5x - 14) = 0$

Let $k = 1$: $x^2 + 5x - 14 = 0$

Let $k = -1$: $-x^2 - 5x + 14 = 0$

Let $k = 2$: $2x^2 + 10x - 28 = 0$

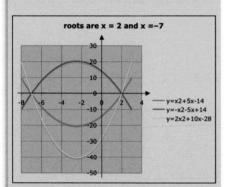

roots are x = 2 and x = −7

y=x2+5x-14
y=-x2-5x+14
y=2x2+10x-28

Example 3

Find the quadratic function whose zeros are $x = 2$ and $x = -7$ and which passes through the point $(1, 16)$.

Response

$x = 2$ or $x = -7$

$\Rightarrow (x - 2)(x + 7) = 0$

$\Rightarrow f(x) = k(x - 2)(x + 7)$

$\Rightarrow f(x) = k(x^2 + 5x - 14)$

We are given that $f(1) = 16$

$\Rightarrow k(1^2 + 5.1 - 14) = 16$

$\Rightarrow -8k = 16$

$\Rightarrow k = -2$

$\Rightarrow f(x) = -2(x^2 + 5x - 14)$

Factor/Remainder Theorem, part 1

Definitions

A polynomial of degree n is a function of the form:
$f(x) = a_0x^0 + a_1x^1 + a_2x^2 + \ldots + a_nx^n$
where $a^0, a^1, \ldots a_n$ are real numbers and $a_n \neq 0$.
For example:

$f(x) = 3x + 1$ is a polynomial of degree 1
… a linear function

$f(x) = 5x^2 + 3x + 1$ is a polynomial of degree 2
… a quadratic function

$f(x) = 2x^3 - 5x^2 + 3x + 1$ is a polynomial of degree 3
… a cubic function

$f(x) = x^4 + 2x^3 + 3x + 1$ is a polynomial of degree 4
… a quartic function
Notice that in this particular example the x^2 term is missing because, as it happens, $a_2 = 0$

Top Tip

When one number is divided by a divisor and the remainder is zero, it stands to reason that the divisor is a factor of the number, and that the answer is another.

Remainder theorem

When a polynomial, $f(x)$ is divided by $x - a$ then the remainder is $f(a)$.

Example
When $2x^3 - 5x^2 + 3x + 1$ is divided by $x - 3$ the remainder is $2.3^3 - 5.3^2 + 3.3 + 1 = 19$.

Synthetic division

This is a quick way of performing a division: e.g. $(2x^3 - 5x^2 + 3x + 1) \div (x - 3)$
When dividing by $(x - 3)$, use a 3 in the manipulations.

Step 1
$$\begin{array}{c|cccc} 3 & 2 & -5 & 3 & 1 \\ & & & & \\ \hline \end{array}$$
set out the coefficients, the a_ns, as shown

Step 2
$$\begin{array}{c|cccc} 3 & 2 & -5 & 3 & 1 \\ & & 6 & & \\ \hline & 2 & & & \end{array}$$
bring down the 1st coefficient and multiply by 3

Step 3
$$\begin{array}{c|cccc} 3 & 2 & -5 & 3 & 1 \\ & & 6 & 3 & \\ \hline & 2 & 1 & & \end{array}$$
add the −5 and 6 and multiply by 3

Step 4
$$\begin{array}{c|cccc} 3 & 2 & -5 & 3 & 1 \\ & & 6 & 3 & 18 \\ \hline & 2 & 1 & 6 & \end{array}$$
add the 3 and 3 and multiply by 3

Step 5
$$\begin{array}{c|cccc} 3 & 2 & -5 & 3 & 1 \\ & & 6 & 3 & 18 \\ \hline & 2 & 1 & 6 & 19 \end{array}$$
add the 1 and 18 to get the remainder of 19

The other parts of the bottom line are the coefficients of the answer to the division.
$(2x^3 - 5x^2 + 3x + 1) \div (x - 3) = 2x^2 + x + 6$ remainder 19.

Note that in step 1, if a term is missing, e.g. $a_n = 0$ for some term, then you should include the 0 in your working.

Top Tip

You may prefer this method – give it a look.

$$(2x^3 - 5x^2 + 3x + 1) \div (x - 3)$$

$$\begin{array}{r} x^2 + x + 6 \\ x - 3 \overline{\smash{\big)}\, 2x^3 - 5x^2 + 3x + 1} \\ \underline{2x^3 - 6x^2} \\ x^2 + 3x + 1 \\ \underline{x^2 - 3x} \\ x + 1 \\ \underline{x - 18} \\ 19 \end{array}$$

Factor theorem

When a polynomial $f(x)$ is divided by $x - a$ and the remainder is zero then $(x - a)$ is a factor of $f(x)$.

Example
Show that $(x - 3)$ is a factor of $(2x^3 - 5x^2 + 3x - 18)$.

Response
Use synthetic division to find the remainder
$$\begin{array}{c|cccc} 3 & 2 & -5 & 3 & -18 \\ & & 6 & 3 & 18 \\ \hline & 2 & 1 & 6 & 0 \end{array}$$

… since the remainder is zero then $x - 3$ is a factor.
… as a bonus, we can see that $2x^2 + x + 6$ is another factor.

Note: Make sure you mention the remainder is zero in the exam… you'll lose marks if you don't show you know it's important.

Factor/Remainder Theorem, part 2

Solving polynomial equations

Top Tip

If a polynomial function equals zero then one of its factors must be zero. To make a polynomial function equal zero, make one of its factors zero.

Example

Solve $x^3 + 2x^2 - 5x - 6 = 0$.

Response

Hunt for a factor. Inspired by the −6 we could try $x \pm 1, x \pm 2, x \pm 3$ considering the 'factors' of −6.

Try $x - 1$
$$\begin{array}{c|cccc} 1 & 1 & 2 & -5 & -6 \\ & & 1 & 3 & -2 \\ \hline & 1 & 3 & -2 & -8 \end{array}$$

Because the remainder is not zero then $x - 1$ is not a factor.

Try $x + 1$
$$\begin{array}{c|cccc} -1 & 1 & 2 & -5 & -6 \\ & & -1 & -1 & 6 \\ \hline & 1 & 1 & -6 & 0 \end{array}$$

Because the remainder is zero, $x - (-1)$ is a factor, i.e. $(x + 1)$ factor $x^2 + x - 6$ is another.

The equation becomes $(x^2 + x - 6)(x + 1) = 0$
The quadratic factorises to give:
$(x - 2)(x + 3)(x + 1) = 0$
$\Rightarrow (x - 2) = 0, (x + 3) = 0$ or $(x + 1) = 0$
$\Rightarrow x = 2, x = -3$ or $x = -1$

Approximate solutions

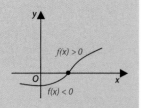

The roots of an equation $f(x) = 0$, can be spotted on a graph as the points where the curve $y = f(x)$ cuts the x-axis.
When the curve is above the x-axis, $f(x) > 0$;
When the curve is below the x-axis, $f(x) < 0$.
When $f(x)$ switches from one state to the other the curve must have crossed the axis.

Top Tip

Find $f(x)$ for various values of x and watch out for a switch of sign.

This knowledge allows us to spot regions where a root exists.

Example
Show that there is a root of the equation $x^3 + 2x^2 - 5x - 5 = 0$ in the region $1 < x < 2$ and find it correct to 2 decimal places.

Response
Consider $f(x) = x^3 + 2x^2 - 5x - 5$.
Using a calculator we find $f(1) = -7$ and $f(2) = 1$: the switch of sign indicates that a root exists between $x = 1$ and $x = 2$, and is closer to 2.
Try $f(1\cdot9) = -0\cdot421$. This replaces the previous negative result, so the root exists between $x = 1\cdot9$ and $x = 2$ and is closer to $1\cdot9$.

Try $f(1\cdot93) = -0\cdot011143$. This replaces the previous negative result, so the root exists between $x = 1\cdot93$ and $x = 2$, and is closer to $1\cdot93$.
Try $f(1\cdot94) = 0\cdot128584$. This replaces the previous positive result, so the root exists between $x = 1\cdot93$ and $x = 1\cdot94$ and is closer to $1\cdot93$.
Try $f(1\cdot931) = 0\cdot00275949$. This replaces the previous positive result, so the root exists between $x = 1\cdot93$ and $x = 1\cdot931$.

When rounded to 2 decimal places both endpoints of the region are $1\cdot93$.
Thus $x = 1\cdot93$ is the required root (to 2 d.p.).

Integration

Things to remember about indices – you'll need to know them for this work:

(i) $x^m = x.x.x. \ldots .x.x$ (m factors)

(ii) $x^m x^n = x^{m+n}$

(iii) $\dfrac{x^m}{x^n} = x^{m-n}$

(iv) $\sqrt[n]{x} = x^{\frac{1}{n}}$

(v) $x^{-n} = \dfrac{1}{x^n}$

(vi) $x^1 = x$

(vii) $x^0 = 1$

Top Tip

You must have a good grasp of the laws of indices.

Definitions

Top Tip

'Add 1 to the power and divide by the new power.'

Definition 1

If $F'(x) = f(x)$ then $\displaystyle\int f(x)\ dx = F(x) + c$ where c is the constant of integration.

Definition 2

If $f(x) = px^n$ then $\displaystyle\int px^n\ dx = \dfrac{px^{n+1}}{n+1} + c$ This is often stated as: 'Add 1 to the power and divide by the new power.'

Definition 3:

If $F'(x) = f(x)$ then $\displaystyle\int_a^b f(x)\ dx = F(b) - F(a)$

Example 1

Integrate $3x^2 - \dfrac{2}{x^3} + \sqrt{x} + 4$

We first make each term look like px^n so we can use the rule in definition 2: $\displaystyle\int 3x^2 - 2x^{-3} + x^{\frac{1}{2}} + 4x^0\ dx$

Now apply the rule to each term to get $= \dfrac{3x^3}{3} - \dfrac{2x^{-2}}{-2} + \dfrac{x^{\frac{3}{2}}}{\frac{3}{2}} + 4x^1 + c$ … don't forget the constant

Finally, tidy up to get: $x^3 + x^{-2} + \frac{2}{3}x^{\frac{3}{2}} + 4x + c$

Example 2

Perform the integration $\displaystyle\int \dfrac{3x^2 + 5x^3 - 4}{x^2}\ dx$

Response

First split the function into separate fractions: $\displaystyle\int \dfrac{3x^2}{x^2} + \dfrac{5x^3}{x^2} - \dfrac{4}{x^2}\ dx$

Next use your knowledge of indices to make each term look like px^n, $\displaystyle\int 3x^0 + 5x^1 - 4x^{-2}\ dx$ …and integrate

$\dfrac{3x^1}{1} + \dfrac{5x^2}{2} - \dfrac{4x^{-1}}{-1} + c$ then tidy up to get: $3x + \frac{5}{2}x^2 + 4x^{-1} + c$

Example 3

Evaluate $\displaystyle\int_{1}^{4} 6x^2 + 4x + \sqrt{x}\ dx$

Response

Prepare the terms $\displaystyle\int_{1}^{4} 6x^2 + 4x + x^{\frac{1}{2}}\ dx$

Integrate, but don't worry about the constant of integration $\left[\dfrac{6x^3}{3} + \dfrac{4x^2}{2} + \dfrac{2x^{\frac{3}{2}}}{3}\right]_{1}^{4}$

Tidy up, and remember you are about to substitute values $\left[2x^3 + 2x^2 + \tfrac{2}{3}\left(\sqrt{x}\right)^3\right]_{1}^{4}$

Substitute $\left[2.4^3 + 2.4^2 + \tfrac{2}{3}\left(\sqrt{4}\right)^3\right] - \left[2.1^3 + 2.1^2 + \tfrac{2}{3}\left(\sqrt{1}\right)^3\right]$

Evaluate: $128 + 32 + {}^{16}/_3 - 2 - 2 - {}^{2}/_3 = 160{}^{2}/_3$

Areas, part 1

Area bounded by the curve $y = f(x)$, the lines $x = a$, $x = b$ and the x-axis

In each case the area, A, is the positive value of the definite integral

$$\int_{a}^{b} f(x)\ dx$$

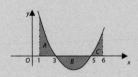

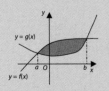

A negative answer is an indication that the area lies below the x-axis.

Example

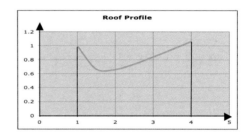

A roof has an unusual design. Its roof-ridge can be modelled by the function

$$f(x) = \frac{1}{x^2} + \sqrt{x} - 1 \qquad \text{where } 1 \le x \le 4.$$

We would want to work out its area to calculate cost. Calculate the area of this side of the roof.

Response

$$\text{Area} = \int_{1}^{4} \frac{1}{x^2} + \sqrt{x} - 1\ dx = \int_{1}^{4} x^{-2} + x^{\frac{1}{2}} - 1\ dx$$

$$= \left[-x^{-1} + \tfrac{2}{3}x^{\frac{3}{2}} - x\right]_{1}^{4} = \left[-4^{-1} + \tfrac{2}{3}.4^{\frac{3}{2}} - 4\right] - \left[-1^{-1} + \tfrac{2}{3}.1^{\frac{3}{2}} - 1\right]$$

$$= -\frac{1}{4} + \frac{16}{3} - 4 + 1 - \frac{2}{3} + 1 = \frac{29}{12}\ \text{units}^2$$

Areas, part 2

Area when the curve crosses the *x*-axis

If you get a negative answer, explain why you 'drop' the negative sign, e.g.
A = –4. This means 4 units² below the *x*-axis.

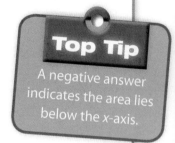

When the curve crosses the *x*-axis you must treat each section as a separate problem, work them out and then add.

Top Tip

A negative answer indicates the area lies below the *x*-axis.

Example

Find the area bound by the function $f(x) = x^2 - 8x + 15$, the lines $x = 1$ and $x = 6$ and the *x*-axis.

Step 1   Find where the function cuts the *x*-axis.
$$x^2 - 8x + 15 = 0$$
$$\Rightarrow (x - 3)(x - 5) = 0$$
$$\Rightarrow x = 3 \text{ or } 5$$

Step 2   You can now make a quick sketch and
identify 3 areas which make up the area required.

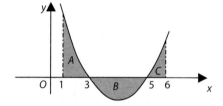

Step 3   Work out the areas

$$A = \int_1^3 x^2 - 8x + 15 \ dx = \left[\frac{x^3}{3} - 4x^2 + 15x \right]_1^3 = 9 - 36 + 45 - \frac{1}{3} + 4 - 15 = 6\frac{2}{3}$$

$$B = \int_3^5 x^2 - 8x + 15 \ dx = \left[\frac{x^3}{3} - 4x^2 + 15x \right]_3^5 = \frac{125}{3} - 100 + 75 - 9 + 36 - 45 = -1\frac{1}{3}$$

$$C = \int_5^6 x^2 - 8x + 15 \ dx = \left[\frac{x^3}{3} - 4x^2 + 15x \right]_5^6 = 72 - 144 + 90 - \frac{125}{3} + 100 - 75 = 1\frac{1}{3}$$

The required area is the sum of these areas (note the area of B is $1^1/_3$ units² below the axis).

Area = $6\frac{2}{3} + 1\frac{1}{3} + 1\frac{1}{3} = 9\frac{1}{3}$ units²

Area between two curves

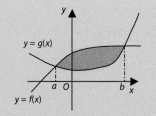

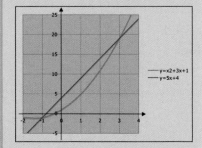

If two curves $y = f(x)$ and $y = g(x)$ intersect at $x = a$ and $x = b$, and if $f(x)$ is above $g(x)$ in that interval then the area trapped between the curves can be worked out by evaluating the definite integral

$$\int_a^b f(x) - g(x) \ dx.$$

Example 1

The diagram shows the line $y = 5x + 4$ and the parabola $y = x^2 + 3x + 1$. Calculate the area enclosed by the line and the curve.

Response

First find the points of intersection:

$x^2 + 3x + 1 = 5x + 4 \Rightarrow x^2 - 2x - 3 = 0$

$\Rightarrow (x + 1)(x - 3) = 0 \Rightarrow x = -1$ or 3

We can now form the appropriate integral:

$$\int_a^b f(x) - g(x) \ dx \text{ becomes } \int_{-1}^{3} (5x + 4) - \left(x^2 + 3x + 1\right) \ dx \ldots \text{ integrate}$$

$$= \int_{-1}^{3} -x^2 + 2x + 3 \ dx = \left[-\frac{x^3}{3} + x^2 + 3x \right]_{-1}^{3} \ldots \text{ perform the substitutions}$$

$$= \left(-\frac{3^3}{3} + 3^2 + 3.3 \right) - \left(-\frac{(-1)^3}{3} + (-1)^2 + 3.(-1) \right) \quad \ldots \text{ and evaluate}$$

$$= -9 + 9 + 9 \ - \ ^1/_3 - 1 + 3 = 10\ ^2/_3 \text{ units}^2$$

Sometimes the context will make the data slightly awkward to deal with.

Example 2

Back at the Squinty Bridge: the arch is modelled by $f(x) = -12x^2 + 144x - 134$ and the road by $g(x) = -x^2 + 12x - 13$. What is the area trapped between the arch and the road?

Response

Step 1   Find where the curves intersect.

$-12x^2 + 144x - 134 = -x^2 + 12x - 13$

$\Rightarrow x^2 - 12x + 11 = 0 \Rightarrow (x - 11)(x - 1) = 0 \Rightarrow x = 11$ or 1

Step 2   The area $= \int_{1}^{11} arch - road \ dx$

$$= \int_{1}^{11} -12x^2 + 144x - 134 + x^2 - 12x + 13 \ dx$$

$$= \left[-\frac{11x^3}{3} + \frac{132x^2}{2} - 121x \right]_{1}^{11}$$

$$= \left(-\frac{14641}{3} + 7986 - 1331 \right) - \left(-\frac{11}{3} + 66 - 121 \right)$$

$$= 1833^1/_3$$

When the curves cross

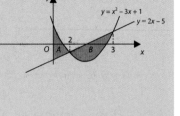

Example

Find the area bounded by the line $y = 2x - 5$ and the curve $y = x^2 - 3x + 1$ in the region $0 \le x \le 3$

Response

Find where the line and curve intersect …

$x^2 - 3x + 1 = 2x - 5$

$\Rightarrow x^2 - 5x + 6 = 0 \Rightarrow (x - 2)(x - 3) = 0 \Rightarrow x = 2 \text{ or } 3$

Make a quick sketch to identify the areas you need.
Area A runs from $0 \le x \le 2$ with the curve on top:

$$A = \int_0^2 (x^2 - 3x + 1) - (2x - 5) \; dx$$

$$A = \int_0^2 x^2 - 5x + 6 \; dx = \left[\frac{x^3}{3} - \frac{5x^2}{2} + 6x \right]_0^2$$

$$A = \left[\frac{2^3}{3} - \frac{5 \cdot 2^2}{2} + 6 \cdot 2 \right] - \left[\frac{0^3}{3} - \frac{5 \cdot 0^2}{2} + 6 \cdot 0 \right] = 4\frac{2}{3}$$

Area B runs from $2 \le x \le 3$ with the line on top:

$$B = \int_2^3 (2x - 5) - (x^2 - 3x + 1) \; dx$$

$$B = \int_2^3 -x^2 + 5x - 6 \; dx = \left[-\frac{x^3}{3} + \frac{5x^2}{2} - 6x \right]_2^3$$

$$B = \left[-\frac{3^3}{3} + \frac{5 \cdot 3^2}{2} - 6 \cdot 3 \right] - \left[-\frac{2^3}{3} + \frac{5 \cdot 2^2}{2} - 6 \cdot 2 \right] = \frac{1}{6}$$

So the total area is $4\frac{2}{3} + \frac{1}{6} = 4\frac{5}{6}$ units^2

Note that when finding the area between curves we don't need to worry about whether parts are above or below the *x*-axis.

Finding functions

Top Tip

This is where you will lose most marks if you forget the constant of integration. Be on guard.

Suppose we know the derivative of a function.

Integrating won't immediately give us our function back: it will give us an infinite supply of possible functions that fit the bill. (Remember the constant of integration.)

However, if we know one point on our function, we can calculate the constant of integration and find our function.

Example

A curve $y = f(x)$ passes through $(1, 3)$. It is known that $f'(x) = 2x + \dfrac{1}{x^2}$

Find the function.

Response

$$f(x) = \int f'(x) \; dx = \int 2x + \frac{1}{x^2} \; dx = \int 2x + x^{-2} \; dx$$

$$\Rightarrow f(x) = x^2 - x^{-1} + c$$

It is given that $f(1) = 3$

The curve passes through $(1, 3)$
$$\Rightarrow f(1) = 1^2 - 1^{-1} + c = 3$$
$$\Rightarrow c = 3$$

$$f(x) = x^2 - x^{-1} + 3 \ldots \quad \text{or} \quad f(x) = x^2 - \frac{1}{x} + 3$$

Trigonometric Equations, part 1

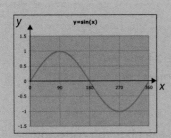 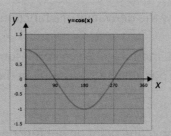 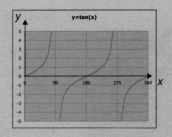

The diagrams show each of the basic trig functions in the region $0 \leq x \leq 360$.
Each function is periodic. It repeats every 360°.
Each graph shows symmetries which can be used to help solve equations.

$\sin x° = \sin(180 - x)°$ ... line symmetry about $x = 90$
$\cos x° = \cos(360 - x)°$ ... line symmetry about $x = 360$
$\tan x° = \tan(180 + x)°$ ... glide symmetry ... shift of 180°

When solving an equation of the form $\text{trig}(x) = a$
    get a first solution using your calculator
    get a second solution by using the symmetries
    get other solutions by adding 360 to known solutions.

When a calculator is not available, or permitted, certain exact values can be deduced from the 'set squares'.

Consider the half-square with sides 1 unit. Pythagoras' Theorem tells us the diagonal is $\sqrt{2}$. From it, we can get the sine, cosine and tangent of 45°.

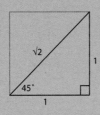

Consider the half-equilateral triangle with sides 2 units. The base is halved to give 1 and Pythagoras' Theorem tells us the altitude is $\sqrt{3}$. From this, we can get the sine, cosine and tangent of 30° and 60°.

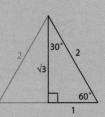

we can see where each ratio is positive.
This helps us get the second angle.

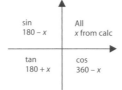

sin $180 - x$	All x from calc
tan $180 + x$	cos $360 - x$

Radian measure

Angles can be measured in degrees. There are 360° in a revolution.
They can also be measured in radians. There are 2π radians in a revolution.
$180° = \pi$ radians.

To turn degrees into radians, multiply by π and divide by 180: $\times \dfrac{\pi}{180}$

To turn radians into degrees, multiply by 180 and divide by π: $\times \dfrac{180}{\pi}$

It is common to leave ϖ in the answer when working in radians.

Examples

(i) $\quad 60° = 60 \times \dfrac{\pi}{180} = \dfrac{60\pi}{180} = \dfrac{\pi}{3}$ radians

(ii) $\quad 150° = 150 \times \dfrac{\pi}{180} = \dfrac{150\pi}{180} = \dfrac{5\pi}{6}$ radians

(iii) $\quad 0.625$ radians $= 0.625 \times \dfrac{180}{\pi} = 35.8°$

This table summarises those exact values you are expected to be able to use in the exam. It is better to learn the above diagrams rather than the table, however.

Degrees	Sine	Cosine	Tangent	Radians
$0°$	0	1	0	0
$30°$	$\frac{1}{2}$	$\frac{\sqrt{3}}{2}$	$\frac{1}{\sqrt{3}}$	$\frac{\pi}{6}$
$45°$	$\frac{1}{\sqrt{2}}$	$\frac{1}{\sqrt{2}}$	1	$\frac{\pi}{4}$
$60°$	$\frac{\sqrt{3}}{2}$	$\frac{1}{2}$	$\sqrt{3}$	$\frac{\pi}{3}$
$90°$	1	0	undefined	$\frac{\pi}{2}$

Simple trigonometric equations at Higher

Example 1

Solve $2\cos x + 5 = 6, 0 \le x < 2\pi$

Response

$2\cos x + 5 = 6$

$\Rightarrow 2\cos x = 1 \quad \Rightarrow \cos x = \frac{1}{2} \Rightarrow x = \frac{\pi}{3}$ (from above table)

or $2\pi - \frac{\pi}{3} = \frac{5\pi}{3}$ (symmetry of cosine)

or $2\pi + \frac{\pi}{3} = \frac{7\pi}{3}$ (from the period)

… but this is more than 2π and outside the search area.

So the solution is : $x = \frac{\pi}{3}$ or $\frac{5\pi}{3}$

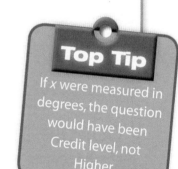

Top Tip

If x were measured in degrees, the question would have been Credit level, not Higher.

Example 2

Solve $2\sin 3x - 5 = 6, 0 \le x < 2\pi \quad \Rightarrow 2\sin 3x = 1$

$\Rightarrow \sin 3x = \frac{1}{2}$

$\Rightarrow 3x = \frac{\pi}{6}$ (from above table)

or $\pi - \frac{\pi}{6} = \frac{5\pi}{6}$ (symmetry of sine)   or $2\pi + \frac{\pi}{6} = \frac{13\pi}{6}$ (from the period)

or $2\pi + \frac{5\pi}{6} = \frac{17\pi}{6}$ (from the period)   or $2\pi + \frac{13\pi}{6} = \frac{25\pi}{6}$ (from the period)

or $2\pi + \frac{17\pi}{6} = \frac{29\pi}{6}$ (from the period)   or $2\pi + \frac{25\pi}{6} = \frac{37\pi}{6}$ (from the period), …

$3x = \frac{\pi}{6}, \frac{5\pi}{6}, \frac{13\pi}{6}, \frac{17\pi}{6}, \frac{25\pi}{6}, \frac{29\pi}{6}, \frac{37\pi}{6}, \quad \dots \Rightarrow x = \frac{\pi}{18}, \frac{5\pi}{18}, \frac{13\pi}{18}, \frac{17\pi}{18}, \frac{25\pi}{18}, \frac{29\pi}{18}, \frac{37\pi}{18}, \dots$

The last one is greater than 2π ($\frac{36\pi}{18}$), and so is omitted from the set of solutions.

Trigonometric Equations, part 2

Compound angle formulae

$\sin(A + B) = \sin A \cos B + \cos A \sin B$
$\sin(A - B) = \sin A \cos B - \cos A \sin B$
$\cos(A + B) = \cos A \cos B - \sin A \sin B$
$\cos(A - B) = \cos A \cos B + \sin A \sin B$

These 4 formulae are given in the exam and are true for any A and B. They are often used in geometric problem solving.

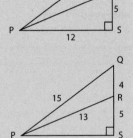

Example
(i)   Find the exact value of: (a) sin QPS; (b) cos RPS
(ii)   Hence find the exact value of sin QPR

Response
We can find both PQ and RP by Pythagoras' theorem

Using elementary trig we can now answer part (i)

(a) $\sin QPS = \dfrac{9}{15} = \dfrac{3}{5}$ ... also, $\cos QPS = \dfrac{12}{15} = \dfrac{4}{5}$

(b) $\cos RPS = \dfrac{12}{13}$ ... also, $\sin RPS = \dfrac{5}{13}$

(ii) $\sin QPR = \sin(QPS - RPS) = \sin QPS \cos RPS - \cos QPS \sin RPS$

$$= \frac{3}{5} \cdot \frac{12}{13} - \frac{4}{5} \cdot \frac{5}{13} = \frac{36}{65} - \frac{20}{65} = \frac{16}{65}$$

Double-angle formulae

$\sin 2A = 2\sin A \cos A$
$\cos 2A = \cos^2 A - \sin^2 A = 2\cos^2 A - 1 = 1 - 2\sin^2 A$
These are commonly used to solve equations involving double angles.

Example 1
Solve $3\sin 2x° + 2\cos x° = 0, 0 \leq x \leq 360$.

Response
$3.2\sin x° \cos x° + 2\cos x° = 0$
$\Rightarrow 2\cos x°(3\sin x° + 1) = 0$
$\Rightarrow 2\cos x° = 0$ or $3\sin x° + 1 = 0$
$\Rightarrow \cos x° = 0$ or $\sin x° = -\frac{1}{3}$
$\Rightarrow x = 90, 360 - 90 = 270$ or $x = -19{\cdot}5, 180 - (-19{\cdot}5), 360 + (-19{\cdot}5)$
$\Rightarrow x = 90, 270, -19{\cdot}5, 199{\cdot}5, 340{\cdot}5$
The negative value is outside the desired region so the set of solutions is
$x = 90, 199{\cdot}5, 270, 340{\cdot}5$

Example 2

Solve $3\cos 2x° - \cos x° + 1 = 0, 0 \leq x \leq 360$

Note: In this case we have a choice of two substitutions … pick the one that matches the 'single-angle term'.
So if the 'single-angle term' contains $\sin x$, use $\cos 2x = 1 - 2\sin^2 x$;
if it contains $\cos x$ … use $\cos 2x = 2\cos^2 x - 1$

Response

$3\cos 2x° - \cos x° + 1 = 0$

$3(2\cos^2 x° - 1) - \cos x° + 1 = 0$

$\Rightarrow 6\cos^2 x° - \cos x° - 2 = 0$

$\Rightarrow (3\cos x - 2)(2\cos x + 1) = 0$

$\Rightarrow \cos x = -\frac{2}{3}$ or $\cos x = -\frac{1}{2}$

$\Rightarrow x = 48{\cdot}2, 360 - 48{\cdot}2 = 311{\cdot}8$ or $x = 120, 360 - 120 = 240$

$\Rightarrow$ set of solutions in the required interval $x = 48{\cdot}2, 120, 240, 311{\cdot}8$

Problem solving

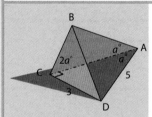

Sometimes, a 2D/3D geometric problem will occur. This will often be solved using your trig knowledge.

Example

ABCD is a pyramid. Its base, ACD, is a right-angled triangle with CD = 3 units and AD = 5 units.

$\angle DAC = \angle CAB$ and $\angle BCA = 2\angle CAB$.

Find the exact value of $\sin(\angle CBA)$.

Response

By considering the base we see that $\sin a° = \frac{3}{5}$ and $\cos a° = \frac{4}{5}$

By considering triangle ABC we see $\angle CBA = [180 - (2a + a)]°$

$\Rightarrow \sin(\angle CBA) = \sin[180 - (2a + a)]° = \sin(2a + a)°$

$= \sin 2a° \cos a° + \cos 2a° \sin a°$

$\sin 2a° = 2\sin a° \cos a° = 2 \cdot \frac{3}{5} \cdot \frac{4}{5} = \frac{24}{25}$ and

$\cos 2a° = \cos^2 a° - \sin^2 a° = \left(\frac{4}{5}\right)^2 - \left(\frac{3}{5}\right)^2 = \frac{16}{25} - \frac{9}{25} = \frac{7}{25}$

$\sin(\angle CBA) = \sin 2a° \cos a° + \cos 2a° \sin a° = \frac{24}{25} \cdot \frac{4}{5} + \frac{7}{25} \cdot \frac{3}{5} = \frac{96}{125} + \frac{21}{125} = \frac{117}{125}$

If no guidance is given, this will become a grade A question – often you will be given hints or asked questions that will lead you through the steps.

The Equation of a Circle, part 1

Centre (a, b), radius r.

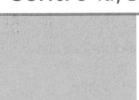

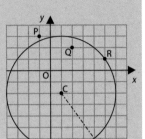

The equation of a circle centre (a, b) and radius r is $(x - a)^2 + (y - b)^2 = r^2$.

Any point that lies on the circumference of the circle satisfies the equation.

If a point is inside the circle then $(x - a)^2 + (y - b)^2 < r^2$.
If a point is outside the circle then $(x - a)^2 + (y - b)^2 > r^2$.

Examples
$(x - 1)^2 + (y + 2)^2 = 25$ is a circle centre $(1, -2)$ and radius $\sqrt{25} = 5$.

Where is the point P$(-1, 3)$?
$(-1 - 1)^2 + (3 + 2)^2 = 29 \ldots 29 > 25$: P is outside the circle.

Where is the point Q$(2, 2)$?
$(2 - 1)^2 + (2 + 2)^2 = 17 \ldots 17 < 25$: Q is inside the circle.

Where is the point R$(5, 1)$?
$(5 - 1)^2 + (1 + 2)^2 = 25 \ldots 25 = 25$: R is on the circle.

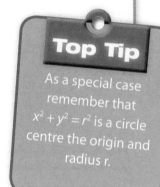

Top Tip

As a special case remember that $x^2 + y^2 = r^2$ is a circle centre the origin and radius r.

Related problems

Given the equation of the circle you should be able to read off the centre and radius *by inspection*.

e.g. $(x - 5)^2 + (y + 1)^2 = 17$ is a circle centre $(5, -1)$ and radius $\sqrt{17}$

Given the centre and radius you should be able to write down the equation.

e.g. a circle centre $(-4, 3)$ and radius 7 has equation $(x + 4)^2 + (y - 3)^2 = 49$

Given the centre and a point on the circumference you should be able to work out the radius (using the distance formula: $r^2 = ((x_2 - x_1)^2 + (y_2 - y_1)^2)$) and then write down the equation of the circle.

e.g. the circle centre $(1, 3)$ passes through $(5, 6)$. Using the distance formula we calculate $r^2 = (5 - 1)^2 + (6 - 3)^2 = 16 + 9 = 25$.

The equation of the circle is $(x - 1)^2 + (y - 3)^2 = 25$

Another form of the equation

The equation $x^2 + y^2 + 2gx + 2fy + c = 0$ will represent a circle if $g^2 + f^2 - c > 0$.
The circle it represents has a centre $(-g, -f)$ and radius $\sqrt{(g^2 + f^2 - c)}$.

Example 1
Find the circle and state its centre and radius.

(i) $x^2 + y^2 - 2x + 4y + 5 = 0$
(ii) $x^2 + y^2 + 2x - 4y + 8 = 0$
(iii) $x^2 + y^2 + 2x + 4y + 1 = 0$

Response
(i)   By inspection, $g = -1, f = 2, c = 5; (-1)^2 + (2)^2 - 5 = 0$, so not a circle
(ii) $g = 1, f = -2, c = 8; (1)^2 + (-2)^2 - 8 = -3 < 0$, so not a circle
(iii) $g = 1, f = 2, c = 1; (1)^2 + (2)^2 - 1 = 4 > 0$, so a circle centre $(-1, -2)$, radius 2.

Example 2
For what values of k does the equation $x^2 + y^2 + 4x + 2ky + 5k - 2 = 0$ represent a circle?

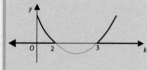

Response
By inspection $g = 2, f = k, c = 5k - 2$
To be a circle $g^2 + f^2 - c > 0$
$\Rightarrow 2^2 + k^2 - (5k - 2) > 0$
$\Rightarrow k^2 - 5k + 6 > 0$
$\Rightarrow (k - 2)(k - 3) > 0$

Making a sketch lets us see that $k < 2$ or $k > 3$

> **Top Tip**
> You can test whether an equation might represent a circle or not by 'completing the square'.

Example 3
By 'completing the square' decide whether the equation $x^2 + y^2 - 2x + 4y + 5 = 0$ represents a circle or not.

Response
Rearrange the equation to bring the x-terms together and the y-terms together.
$\underline{x^2 - 2x} + \underline{y^2 + 4y} + 5 = 0$, completing the square on the underlined parts,
$(x - 1)^2 - 1^2 + (y + 2)^2 - 2^2 + 5 = 0$
$(x - 1)^2 + (y + 2)^2 = 0$
This would suggest a circle with zero radius.
The equation cannot represent a circle.

The Equation of a Circle, part 2

The line and the circle

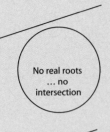

No real roots ... no intersection

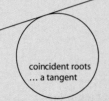

coincident roots ... a tangent

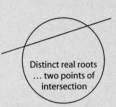

Distinct real roots ... two points of intersection

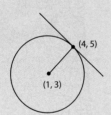

(4, 5)

(1, 3)

We find where the line $y = ax + b$ meets the circle $x^2 + y^2 + 2gx + 2fy + c = 0$ by replacing y by $ax + b$ in the equation of the circle.

This will produce a quadratic equation to solve.

Example 1

Find where the line $y = x + 1$ cuts the circle $x^2 + y^2 - 2x - 6y - 15 = 0$.

Response

Substitute $x + 1$ for y in the equation of the circle.

$x^2 + (x + 1)^2 - 2x - 6(x + 1) - 15 = 0$
$\Rightarrow x^2 + x^2 + 2x + 1 - 2x - 6x - 6 - 15 = 0$
$\Rightarrow 2x^2 - 6x - 20 = 0$
$\Rightarrow x^2 - 3x - 10 = 0$
$\Rightarrow (x + 2)(x - 5) = 0$
$\Rightarrow x = -2$ or 5
$\Rightarrow y = -1$ or 6

The points of intersection are $(-2, -1)$ and $(5, 6)$

Example 2

Show that the line $4x + y - 10 = 0$ is a tangent to the circle $x^2 + y^2 + 6x - 10y + 17 = 0$.

Response

Rearranging the equation of the line we get: $y = -4x + 10$
Substituting into circle: $x^2 + (-4x + 10)^2 + 6x - 10(-4x + 10) + 17 = 0$
$\Rightarrow x^2 + 16x^2 - 80x + 100 + 6x + 40x - 100 + 17 = 0$
$\Rightarrow 17x^2 - 34x + 17 = 0$
$\Rightarrow x^2 - 2x + 1 = 0$
$\Rightarrow (x - 1)(x - 1) = 0$
$\Rightarrow x = 1$ or 1 (coincident roots)
$\Rightarrow y = 6$ or 6

Line contacts circle at only one point $(1, 6)$.
So line is a tangent to the circle.

Example 3

Find the equation of the tangent to the circle $x^2 + y^2 - 2x - 6y - 3 = 0$ at the point $(4, 5)$.

Response

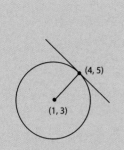

By inspection the centre of the circle is $(1, 3)$.

The gradient of the radius to the point of tangency is $\dfrac{5-3}{4-1} = \dfrac{2}{3}$

So the gradient of the tangent (which is at right angles to this radius) is $\dfrac{-3}{2}$

The equation of the tangent: $(y - 5) = \dfrac{-3}{2}(x - 4)$

This tidies up to $3x + 2y - 22 = 0$.

Problem solving

Example

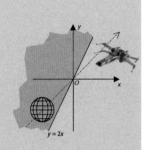

In the movie 'Star Battles', a computer graphic of the 'Doomstar' emerging from the 'Dark Zone' appears every few minutes to heighten the tension.

The rim of the Doomstar is represented by a circle $(x - k)^2 + (y - k)^2 = 5$.

The edge of the Dark Zone is represented by the line $y = 2x$.

(a) For what value of k will the Doomstar start to emerge from the Dark Zone?

(b) When will it be fully exposed?

Response

The question is asking, 'For what value of k is the line a tangent to the circle?'

Substitute $2x$ for y in the circle: $(x - k)^2 + (2x - k)^2 = 5$

$\Rightarrow x^2 - 2kx + k^2 + 4x^2 - 4kx + k^2 - 5 = 0$

$\Rightarrow 5x^2 - 6kx + (2k^2 - 5) = 0$

For tangency we need coincident roots: $b^2 - 4ac = 0$

$\Rightarrow 36k^2 - 4.5.(2k^2 - 5) = 0$

$\Rightarrow -4k^2 + 100 = 0$

$\Rightarrow k^2 = 25$

$\Rightarrow k = 5$ or -5

(a) The diagram shows that when the centre is $(-5, -5)$, the star starts to emerge

(b) … and when the centre is $(5, 5)$ it is fully exposed.

The Equation of a Circle, part 3

Two circles

By considering a smaller circle passing in front of a larger we can examine all the cases: Let d be the distance between the centres; let r_1 be the radius of the smaller circle and r_2 the radius of the larger.

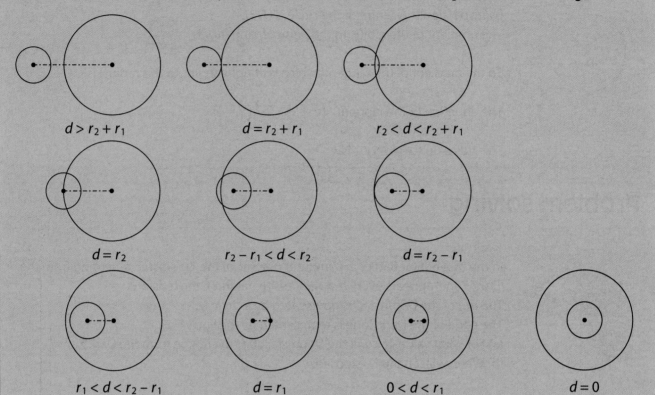

$d > r_2 + r_1$ $d = r_2 + r_1$ $r_2 < d < r_2 + r_1$

$d = r_2$ $r_2 - r_1 < d < r_2$ $d = r_2 - r_1$

$r_1 < d < r_2 - r_1$ $d = r_1$ $0 < d < r_1$ $d = 0$

If $d = r_2 + r_1$, the smaller circle lies outside the larger but touches it.
If $d = r_2 - r_1$, the smaller circle lies inside the larger but touches it.

If $r_2 - r_1 < d < r_2 + r_1$, the circles intersect at two distinct points.

Example
Prove that the circles with equations $x^2 + y^2 - 4x + 2y + 1 = 0$ and
$x^2 + y^2 - 10x - 6y + 18 = 0$ intersect at two points.

Response
By inspection
     the first circle has centre $(2, -1)$ and radius $\sqrt{(-2)^2 + 1^2 - 1} = 2$

     the second circle has centre $(5, 3)$ and radius $\sqrt{(-5)^2 + (-3)^2 - 18} = 4$

     The distance, d, between the centres is $\sqrt{(5 - 2)^2 + (3 - (-1))^2} = 5$

$r_2 - r_1 = 4 - 2 = 2; r_2 + r_1 = 4 + 2 = 6$
$2 < 5 < 6$
Since $r_2 - r_1 < d < r_2 + r_1$ the circles intersect at two distinct points.

This completes Maths 2.
For the unit test, you will be expected to show that you can perform the following under exam conditions. The SQA have published this list in their conditions and arrangements.

OUTCOME 1

Use the Factor/Remainder Theorem and apply quadratic theory.

Performance criteria
a) Apply the Factor/Remainder Theorem to a polynomial function.
b) Determine the nature of the roots of a quadratic equation using the discriminant.

OUTCOME 2

Use basic integration.

Performance criteria
a) Integrate functions reducible to the sums of powers of x (definite and indefinite).
b) Find the area between a curve and the x-axis using integration.
c) Find the area between two curves using integration.

OUTCOME 3

Solve trigonometric equations and apply trigonometric formulae.

Performance criteria
a) Solve a trigonometric equation in a given interval.
b) Apply a trigonometric formula (addition formula) in the solution of a geometric problem.
c) Solve a trigonometric equation involving an addition formula in a given interval.

OUTCOME 4

Use the equation of the circle.

Performance criteria
a) Given the centre (a, b) and radius r, find the equation of the circle in the form: $(x - a)^2 + (y - b)^2 = r^2$.
b) Find the radius and centre of a circle given the equation in the form: $x^2 + y^2 + 2gx + 2fy + c = 0$.
c) Determine whether a given line is a tangent to a given circle.
d) Determine the equation of the tangent to a given circle given the point of contact.

Top Tip

Practice tests suitable for preparing for the unit test can be downloaded from www.leckieandleckie.co.uk.

Vectors (the basics)

Definitions

A **vector** is a quantity with magnitude and direction.
A **scalar** is a quantity with magnitude only.

The **magnitude** of the vector is its size.
The magnitude of vector u is denoted by $|u|$

Note that a vector is often denoted by a lower-case letter, italics and bold.
When hand-written, a short wavy line is usually put under the letter to denote bold i.e. $\underset{\sim}{u}$

A vector is usually represented by a **directed line segment**.
The **length** of the segment is proportional to the magnitude, and the **direction** is indicated by an arrowhead.

A vector can be described by giving its **components**, the shift in the x-direction, the y-direction and the z-direction, made going from tail to nose in the representative.

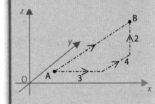

e.g ... $\overrightarrow{AB} = \begin{pmatrix} 3 \\ 4 \\ 2 \end{pmatrix}$. Note the arrow above the AB indicating the direction.

The **position vector** of a point is the displacement of the point from the origin.

The point $A(x_a, y_a, z_a)$ has a position vector $\begin{pmatrix} x_a \\ y_a \\ z_a \end{pmatrix}$.

A **unit vector** has a magnitude of 1 unit.

The **basis vectors**, i, j and k are unit vectors in the x, y and z directions respectively.

$$i = \begin{pmatrix} 1 \\ 0 \\ 0 \end{pmatrix}; j = \begin{pmatrix} 0 \\ 1 \\ 0 \end{pmatrix}; k = \begin{pmatrix} 0 \\ 0 \\ 1 \end{pmatrix}$$

Other vectors can be expressed in terms of the basis vectors.

e.g. $\overrightarrow{AB} = \begin{pmatrix} 3 \\ 4 \\ 2 \end{pmatrix}$

$$= \begin{pmatrix} 3 \\ 0 \\ 0 \end{pmatrix} + \begin{pmatrix} 0 \\ 4 \\ 0 \end{pmatrix} + \begin{pmatrix} 0 \\ 0 \\ 2 \end{pmatrix}$$

$$= 3\begin{pmatrix} 1 \\ 0 \\ 0 \end{pmatrix} + 4\begin{pmatrix} 0 \\ 1 \\ 0 \end{pmatrix} + 2\begin{pmatrix} 0 \\ 0 \\ 1 \end{pmatrix}$$

$$= 3i + 4j + 2k$$

Manipulation

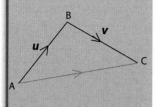

Addition

If $\boldsymbol{u}$ is represented by $\overrightarrow{AB}$
and $\boldsymbol{v}$ is represented by $\overrightarrow{BC}$
then $\boldsymbol{u} + \boldsymbol{v}$ is represented by $\overrightarrow{AC}$.

Note that the vectors to be added are placed nose-to-tail.
Add vectors in component form by adding corresponding components.

Example

Given $\boldsymbol{u} = \begin{pmatrix} -1 \\ 2 \\ 5 \end{pmatrix}$ and $\boldsymbol{v} = \begin{pmatrix} 2 \\ -3 \\ 4 \end{pmatrix}$ then $\boldsymbol{u} + \boldsymbol{v} = \begin{pmatrix} -1 \\ 2 \\ 5 \end{pmatrix} + \begin{pmatrix} 2 \\ -3 \\ 4 \end{pmatrix} = \begin{pmatrix} -1+2 \\ 2+(-3) \\ 5+4 \end{pmatrix} = \begin{pmatrix} 1 \\ -1 \\ 9 \end{pmatrix}$

You may prefer to use basis vectors when performing manipulations. In this case the above becomes:

$\boldsymbol{u} = -\boldsymbol{i} + 2\boldsymbol{j} + 5\boldsymbol{k}$; $\boldsymbol{v} = 2\boldsymbol{i} - 3\boldsymbol{j} + 4\boldsymbol{k}$ then $\boldsymbol{u} + \boldsymbol{v} = -\boldsymbol{i} + 2\boldsymbol{j} + 5\boldsymbol{k} + 2\boldsymbol{i} - 3\boldsymbol{j} + 4\boldsymbol{k} = \boldsymbol{i} - \boldsymbol{j} + 9\boldsymbol{k}$

Subtraction is the inverse of addition.

Example

Given $\boldsymbol{u} = \begin{pmatrix} -1 \\ 2 \\ 5 \end{pmatrix}$ and $\boldsymbol{v} = \begin{pmatrix} 2 \\ -3 \\ 4 \end{pmatrix}$ then $\boldsymbol{u} - \boldsymbol{v} = \begin{pmatrix} -1 \\ 2 \\ 5 \end{pmatrix} - \begin{pmatrix} 2 \\ -3 \\ 4 \end{pmatrix} = \begin{pmatrix} -1-2 \\ 2-(-3) \\ 5-4 \end{pmatrix} = \begin{pmatrix} -3 \\ 5 \\ 1 \end{pmatrix}$

Using basis vectors:

$\boldsymbol{u} = -\boldsymbol{i} + 2\boldsymbol{j} + 5\boldsymbol{k}$; $\boldsymbol{v} = 2\boldsymbol{i} - 3\boldsymbol{j} + 4\boldsymbol{k}$ then $\boldsymbol{u} - \boldsymbol{v} = -\boldsymbol{i} + 2\boldsymbol{j} + 5\boldsymbol{k} - 2\boldsymbol{i} + 3\boldsymbol{j} - 4\boldsymbol{k} = -3\boldsymbol{i} + 5\boldsymbol{j} + \boldsymbol{k}$

Position vectors

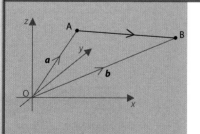

The position vector of A is $\boldsymbol{a}$; the position vector of B is $\boldsymbol{b}$.

By the definition of vector addition, $\boldsymbol{a} + \overrightarrow{AB} = \boldsymbol{b}$.
This gives the important result $\overrightarrow{AB} = \boldsymbol{b} - \boldsymbol{a}$

Top Tip

$\overrightarrow{AB} = \boldsymbol{b} - \boldsymbol{a}$ is the most used fact in this area of work.

This allows us to express the vector represented by a directed line in component form quickly when the end-points are known.

Example

Given the points A(3, –1, 4) and B(5, 2, –1) then

$$\overrightarrow{AB} = \boldsymbol{b} - \boldsymbol{a} = \begin{pmatrix} 5 \\ 2 \\ -1 \end{pmatrix} - \begin{pmatrix} 3 \\ -1 \\ 4 \end{pmatrix} = \begin{pmatrix} 5-3 \\ 2-(-1) \\ -1-4 \end{pmatrix} = \begin{pmatrix} 2 \\ 3 \\ -5 \end{pmatrix}$$

Multiplication by a scalar

This is considered repeated addition.

Example

If $u = \begin{pmatrix} 3 \\ -2 \\ 4 \end{pmatrix}$ then $6u = 6\begin{pmatrix} 3 \\ -2 \\ 4 \end{pmatrix} = \begin{pmatrix} 6 \times 3 \\ 6 \times -2 \\ 6 \times 4 \end{pmatrix} = \begin{pmatrix} 18 \\ -12 \\ 24 \end{pmatrix}$

Since it is repeated addition of the one vector to itself, the product will be parallel to the original vector.

An important result:
If $u = kv$ where k is any scalar then u is parallel to v.
The converse is also true: If u is parallel to v then $u = kv$ for some number k.

Collinear points

The above result can be used to prove line segments are parallel or, if the lines share a common point, parts of the same line.
Given three points, you can test if they lie on the same line, i.e. if they are collinear.

Example

The surveyor at a tunnelling project identifies three target spots to dig towards. Relative to a suitable set of axes their coordinates are A(1, 2, 7), B(3, 8, 3) and C(6, 17, −3).
Show that these three points are collinear.

Response

$\overrightarrow{AC} = c - a = \begin{pmatrix} 6 \\ 17 \\ -3 \end{pmatrix} - \begin{pmatrix} 1 \\ 2 \\ 7 \end{pmatrix} = \begin{pmatrix} 5 \\ 15 \\ -10 \end{pmatrix}$

$\overrightarrow{AB} = b - a = \begin{pmatrix} 3 \\ 8 \\ 3 \end{pmatrix} - \begin{pmatrix} 1 \\ 2 \\ 7 \end{pmatrix} = \begin{pmatrix} 2 \\ 6 \\ -4 \end{pmatrix}$

Note that $2 \div 5 = 6 \div 15 = -4 \div (-10) = 0.4$
$\Rightarrow \overrightarrow{AB} = 0.4\,\overrightarrow{AC}$ and so AB is parallel to AC.

However, A is a common point to both line segments.
So A, B and C all lie on the one straight line.

Top Tip

It is essential that the common point gets a mention or a mark will be lost.

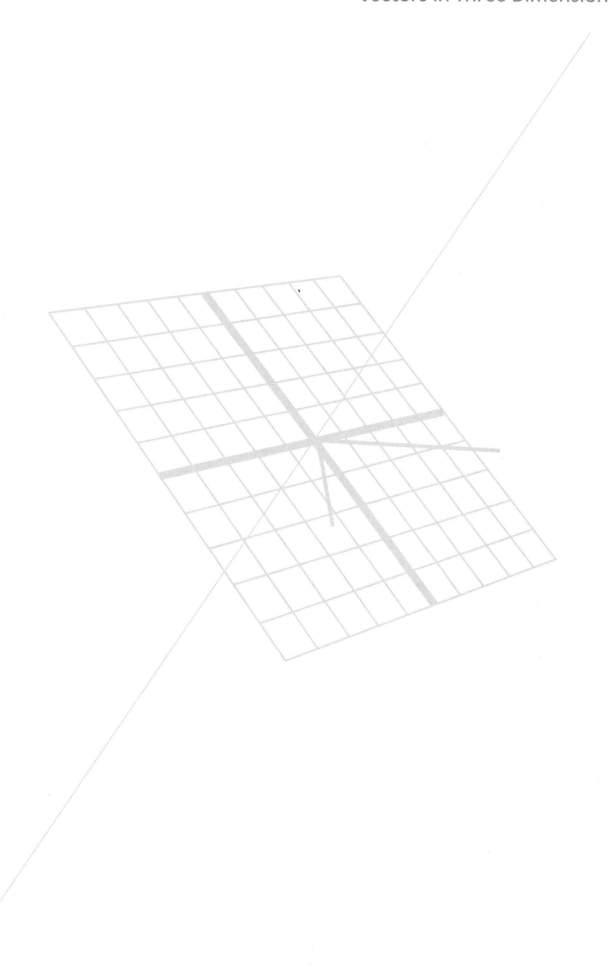

Magnitude, Equality and Sectioning Lines

Magnitude

The distance between A(x_a, y_a, z_a) and B(x_b, y_b, z_b) is

$$AB = \sqrt{(x_b - x_a)^2 + (y_b - y_a)^2 + (z_b - z_a)^2}$$

If a vector $\boldsymbol{a} = \begin{pmatrix} x_a \\ y_a \\ z_a \end{pmatrix}$ then its **magnitude** can be calculated by $|\boldsymbol{a}| = \sqrt{x_a^2 + y_a^2 + z_a^2}$

A reasonable exam question may ask you to find a unit vector parallel to some given vector.

Example
Find a unit vector, $\boldsymbol{v}$, parallel to $\boldsymbol{u} = \begin{pmatrix} 3 \\ -4 \\ 12 \end{pmatrix}$.

Response
First, find the magnitude of the given vector: $|\boldsymbol{u}| = \sqrt{3^2 + (-4)^2 + 12^2} = 13$.

Now multiply $\boldsymbol{u}$ by $\dfrac{1}{|\boldsymbol{u}|}$ ... $\boldsymbol{v} = \dfrac{1}{13}\begin{pmatrix} 3 \\ -4 \\ 12 \end{pmatrix} = \begin{pmatrix} 3/13 \\ -4/13 \\ 12/13 \end{pmatrix}$

It is parallel to $\boldsymbol{u}$ because $\boldsymbol{v} = \frac{1}{13}\boldsymbol{u}$. Its magnitude $= |\boldsymbol{v}| = \frac{1}{13}|\boldsymbol{u}| = \frac{1}{13} \times 13 = 1$.

Equality

Two vectors are equal if, and only if, the corresponding components are equal.

e.g. $\begin{pmatrix} a \\ b \\ c \end{pmatrix} = \begin{pmatrix} d \\ e \\ f \end{pmatrix} \Leftrightarrow a = d, b = e, c = f$

Example
Find x and z such that the vectors $\begin{pmatrix} 1 \\ 2 \\ 3 \end{pmatrix}$ and $\begin{pmatrix} x+1 \\ 6 \\ z+4 \end{pmatrix}$ are parallel.

Response
Because they are parallel we know $k\begin{pmatrix} 1 \\ 2 \\ 3 \end{pmatrix} = \begin{pmatrix} x+1 \\ 6 \\ z+4 \end{pmatrix}$, where k is a constant.

So $\begin{pmatrix} k \\ 2k \\ 3k \end{pmatrix} = \begin{pmatrix} x+1 \\ 6 \\ z+4 \end{pmatrix}$

We can form the system of equations:

$k = x + 1$ ...①

$2k = 6$ ... ②

$3k = z + 4$ ... ③

Equation ② tells us that $k = 3$

Equation ① becomes $3 = x + 1 \Rightarrow x = 2$

Equation ③ becomes $9 = z + 4 \Rightarrow z = 5$

Sectioning lines

When we say a point P divides a line AB in the ratio $m:n$, we mean

$$\frac{\left|\overrightarrow{AP}\right|}{\left|\overrightarrow{PB}\right|} = \frac{m}{n}$$ and we can write $\overrightarrow{AP} = \frac{m}{n}\overrightarrow{PB}$ or indeed $n\overrightarrow{AP} = m\overrightarrow{PB}$

Internally

P might lie between A and B in which case we say P divides AB internally.

Example

A is the point (0, 3, 1) and B is (10, 23, –9).

Find the point, P, which divides the line AB internally in the ratio 3:2.

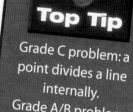

Response

$$\frac{\left|\overrightarrow{AP}\right|}{\left|\overrightarrow{PB}\right|} = \frac{3}{2} \Rightarrow 2\overrightarrow{AP} = 3\overrightarrow{PB}$$

$\Rightarrow \quad 2(\boldsymbol{p} - \boldsymbol{a}) = 3(\boldsymbol{b} - \boldsymbol{p})$ ... $\boldsymbol{a}$, $\boldsymbol{b}$ and $\boldsymbol{p}$ are the position vectors of A, B, P respectively

$\Rightarrow \quad 2\boldsymbol{p} - 2\boldsymbol{a} = 3\boldsymbol{b} - 3\boldsymbol{p}$

$\Rightarrow \quad 5\boldsymbol{p} = 3\boldsymbol{b} + 2\boldsymbol{a}$

$\Rightarrow \quad \boldsymbol{p} = \dfrac{3\boldsymbol{b} + 2\boldsymbol{a}}{5}$

$$\Rightarrow \quad \boldsymbol{p} = \frac{3\begin{pmatrix}10\\23\\-9\end{pmatrix} + 2\begin{pmatrix}0\\3\\1\end{pmatrix}}{5} = \frac{\begin{pmatrix}30+0\\69+6\\-27+2\end{pmatrix}}{5} = \frac{\begin{pmatrix}30\\75\\-25\end{pmatrix}}{5} = \begin{pmatrix}6\\15\\-5\end{pmatrix}$$

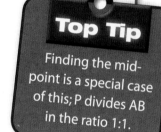

So P is the point (6, 15, –5)

Scalar Product

Definition and properties

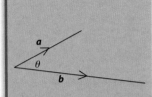

(i) The **scalar product**, *a.b*, of two vectors *a* and *b* is defined as *a.b* = |*a*||*b*|cosθ, where θ is the angle between representatives of the vectors when situated tail-to-tail.

As its name suggests, it is not a vector but a scalar.

(ii) Note that *a.b* = *b.a* and *a.*(*b* + *c*) = *a.b* + *a.c*

(iii) Since cos 90° = 0 then, if two vectors are perpendicular to each other, their scalar product is zero.

The converse holds as long as the vectors are not zero vectors.

i.e. if |*a*|, |*b*| ≠ 0 and *a.b* = 0 then *a* and *b* are perpendicular.

Consider the basis vectors (which are at right angles to each other):
i.j = *j.k* = *k.i* = 0

Top Tip

It is critical that the vectors are tail-to-tail. Drawing a sketch is a good way of making sure this is the case.

Example

The vectors $\begin{pmatrix} 3 \\ -2 \\ 7 \end{pmatrix}$ and $\begin{pmatrix} k \\ 2k \\ k+3 \end{pmatrix}$ are mutually perpendicular.

What is the value of *k*?

Top Tip

If the scalar product is zero then the two vectors are mutually perpendicular and vice versa.

Response

$$\begin{pmatrix} 3 \\ -2 \\ 7 \end{pmatrix} \cdot \begin{pmatrix} k \\ 2k \\ k+3 \end{pmatrix} = 3k - 4k + 7k + 21 = 6k + 21$$

Because the vectors are at right angles, this is equal to zero.
$6k + 21 = 0$
$\Rightarrow k = -3.5$

(iv) Because cos 0° = 1 then *a.a* = |*a*| |*a*| cos 0° = |*a*|²

Consider the basis vectors (each of which has a magnitude of 1):
i.i = *j.j* = *k.k* = 1² = 1

(v) If $a = \begin{pmatrix} a_1 \\ a_2 \\ a_3 \end{pmatrix}$ and $b = \begin{pmatrix} b_1 \\ b_2 \\ b_3 \end{pmatrix}$ then *a.b* = $a_1b_1 + a_2b_2 + a_3b_3$

Example

Use basis vectors to prove $\mathbf{a.b} = a_1b_1 + a_2b_2 + a_3b_3$

Response

$$\mathbf{a.b} = (a_1\mathbf{i} + a_2\mathbf{j} + a_3\mathbf{k}).(b_1\mathbf{i} + b_2\mathbf{j} + b_3\mathbf{k})$$
$$= a_1b_1\mathbf{i.i} + a_1b_2\mathbf{i.j} + a_1b_3\mathbf{i.k} + a_2b_1\mathbf{j.i} + a_2b_2\mathbf{j.j} + a_2b_3\mathbf{j.k} + a_3b_1\mathbf{k.i} + a_3b_2\mathbf{k.j} + a_3b_3\mathbf{k.k}$$

Now $\mathbf{i.i} = \mathbf{j.j} = \mathbf{k.k} = 1$ and $\mathbf{i.j} = \mathbf{j.k} = \mathbf{k.i} = 0$

$\Rightarrow \mathbf{a.b} = a_1b_1 + a_2b_2 + a_3b_3$

(vi) Using the above, we can calculate the angle between two vectors in three-dimensional space.

$$\mathbf{a.b} = |\mathbf{a}||\mathbf{b}|\cos\theta \Rightarrow \cos\theta = \frac{\mathbf{a.b}}{|\mathbf{a}||\mathbf{b}|}$$

Example

Two communication satellites, Beacon and Calypso, are parked above the Earth. Relative to a suitable set of axes their coordinates are
B(3, 7, –2) and C(1, –5, 6).
On Earth a receiving station has coordinates A(0, 1, 2).
Calculate the angle of separation of the satellites as measured from the station i.e. $\angle$BAC.

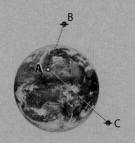

Response

$$\overrightarrow{AB} = \mathbf{b} - \mathbf{a} = \begin{pmatrix} 3 \\ 7 \\ -2 \end{pmatrix} - \begin{pmatrix} 0 \\ 1 \\ 2 \end{pmatrix} = \begin{pmatrix} 3 \\ 6 \\ -4 \end{pmatrix}$$

$$\left|\overrightarrow{AB}\right| = \sqrt{3^2 + 6^2 + (-4)^2} = \sqrt{61}$$

$$\overrightarrow{AC} = \mathbf{c} - \mathbf{a} = \begin{pmatrix} 1 \\ -5 \\ 6 \end{pmatrix} - \begin{pmatrix} 0 \\ 1 \\ 2 \end{pmatrix} = \begin{pmatrix} 1 \\ -6 \\ 4 \end{pmatrix}$$

$$\left|\overrightarrow{AC}\right| = \sqrt{1^2 + (-6)^2 + 4^2} = \sqrt{53}$$

$$\overrightarrow{AB}.\overrightarrow{AC} = \begin{pmatrix} 3 \\ 6 \\ -4 \end{pmatrix}.\begin{pmatrix} 1 \\ -6 \\ 4 \end{pmatrix} = 3.1 + 6.(-6) + (-4).4 = -49$$

Using $\cos\theta = \dfrac{\mathbf{a.b}}{|\mathbf{a}||\mathbf{b}|}$, $\cos\angle BAC = \dfrac{-49}{\sqrt{61}.\sqrt{53}} = -0.861774$

so $\angle BAC = 149.5°$

Differentiation

Trigonometric functions

If we work in **radians**, differentiating trigonometric functions becomes easy.

If $f(x) = \sin x$ then $f'(x) = \cos x$

If $f(x) = \cos x$ then $f'(x) = -\sin x$

Everything you learned in Unit 1 about differentiation can be used here.

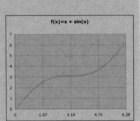

Example 1

Prove that the function, $f(x) = x + \sin x$ is an increasing function.

Response

$f(x) = x + \sin x$

$\Rightarrow f'(x) = 1 + \cos x$. Since $-1 \le \cos x \le 1$

we have $1 - 1 \le 1 + \cos x \le 1 + 1$

i.e. $0 \le f'(x) \le 2$.

The derivative is never negative, so the function is never decreasing.

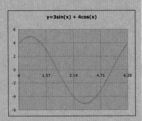

Example 2

Find the maximum value of $3 \sin x + 4 \cos x$

Response

$f(x) = 3 \sin x + 4 \cos x$

$\Rightarrow f'(x) = 3 \cos x - 4 \sin x$. At stationary points $f'(x) = 0$

$\Rightarrow 3 \cos x - 4 \sin x = 0$

$\Rightarrow \tan x = \frac{3}{4}$

$x = 0\cdot 64350111$ (Remember we work in radians when we work in calculus.)

The sketch shows us that this is when a maximum occurs

$\Rightarrow$ the maximum value is $3 \sin (0\cdot 64350111) + 4 \cos (0\cdot 64350111) = 5$

Example 3

A right-angled triangle has a hypotenuse of 1 unit. One of the shorter sides makes an angle of x radians with the hypotenuse. What value of x will maximise the perimeter of the triangle?

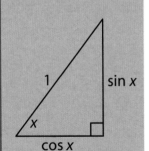

Response

A quick sketch and some elementary trig will lead to the fact that the perimeter, $P = 1 + \cos x + \sin x$, $0 < x < \frac{\pi}{2}$

$\dfrac{dP}{dx} = -\sin x + \cos x$

The derivative is zero at stationary points.

$-\sin x + \cos x = 0$

$\Rightarrow \tan x = 1$

$\Rightarrow x = \frac{\pi}{4}$

x	$\rightarrow$	$\frac{\pi}{4}$	$\rightarrow$
$\frac{dP}{dx}$	+	0	−
	/	−	\

This is a maximum turning point.

The corresponding perimeter is

$1 + \cos(\frac{\pi}{4}) + \sin(\frac{\pi}{4})$

$= 1 + \frac{1}{\sqrt{2}} + \frac{1}{\sqrt{2}}$

$= 1 + \frac{2}{\sqrt{2}}$

$= 1 + \sqrt{2}$

Top Tip

The following are given in the exam paper:

If $f(x) = \sin ax$ then $f'(x) = a\cos ax$

If $f(x) = \cos ax$ then $f'(x) = -a\sin ax$

... but they are worth learning.

The chain rule

We can differentiate composite functions using this rule.

If $h(x) = f(g(x))$ then $h'(x) = f'(g(x)) \times g'(x)$

Often this is remembered as:
'Differentiate outside and multiply by the derivative of the inside.'

Using Leibniz notation it is often quoted as $\dfrac{dy}{dx} = \dfrac{dy}{du} \cdot \dfrac{du}{dx}$
where u is the 'inside' function.

Example 1
Differentiate $(3x + 2)^5$

Response
Consider the 'outside function' as $(---)^5$ and the 'inside function' as $3x + 2$
$$\frac{dy}{dx} = 5(3x + 2)^4 \times 3 = 15(3x + 2)^4$$

We would normally use implicit multiplication rather than use the $\times$ symbol.

i.e. $\dfrac{dy}{dx} = 5(3x + 2)^4 \cdot 3 = 15(3x + 2)^4$

Example 2
$y = \sin(3x + 2)$. Find $\dfrac{dy}{dx}$

Response
Consider the 'outside function' as $\sin(---)$ and the 'inside function' as $3x + 2$

$$\frac{dy}{dx} = \cos(3x + 2) \cdot 3 = 3\cos(3x + 2)$$

Example 3
Find $f'(x)$ when $f(x) = \sqrt{\sin 3x}$

Response
$f(x) = \sqrt{\sin 3x} = (\sin 3x)^{\frac{1}{2}}$

Consider the 'outside function' as $(---)^{\frac{1}{2}}$ and the 'inside function' as $\sin 3x$

$$\frac{dy}{dx} = \frac{1}{2}(\sin 3x)^{-\frac{1}{2}} \cdot \frac{d}{dx}(\sin 3x)$$

We have to use the chain rule again:

$$\frac{dy}{dx} = \frac{1}{2}(\sin 3x)^{-\frac{1}{2}} \cdot (\cos 3x) \cdot 3$$

Top Tip

'Differentiate outside and multiply by the derivative of the inside.' This increases the number of functions you can differentiate; the types of problems that you use differentiation to solve are still the same.

Integration

Trigonometric functions

If we work in radians, integrating trigonometric functions becomes easy.

$$\int \sin x \; dx = -\cos x + c \quad \Big| \quad \int \cos x \; dx = \sin x + c$$

Everything you learned in Unit 2 about integration can be used here.

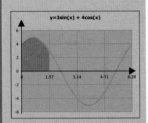

Example 1

Find the area that lies between the curve $y = 3\sin x + 4\cos x$, the lines $x = 0, x = \pi/2$ and the x-axis.

Response

$$\text{The required area} = \int_0^{\frac{\pi}{2}} 3\sin x + 4\cos x \; dx = \left[-3\cos x + 4\sin x\right]_0^{\frac{\pi}{2}}$$

$$= \left(-3\cos \tfrac{\pi}{2} + 4\sin \tfrac{\pi}{2}\right) - \left(-3\cos 0 + 4\sin 0\right)$$

$$= 4 - (-3)$$

$$= 7$$

The area is 7 units^2.

Example 2

As a water wheel turns, the velocity, v m/s of a point, P, on its circumference towards the end wall is given by $v = -3\sin t$ where t is the time in seconds since observations began.

It is known that when $t = 0$ seconds, $s = 10$ metres. Find an expression for the distance, s metres, from P to the end wall.

Response

You should remember about rates of change that

$$v = \frac{ds}{dt} \quad \text{and hence} \quad s = \int v \; dt$$

So $s = \int -3\sin t \; dt = 3\cos t + c$

When $t = 0$ seconds, $s = 10$ metres.

$$\Rightarrow 10 = 3\cos 0 + c = 3 + c$$

$$\Rightarrow c = 7$$

Thus $s = 3\cos t + 7$

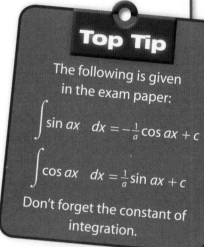

Top Tip

The following is given in the exam paper:

$$\int \sin ax \; dx = -\tfrac{1}{a}\cos ax + c$$

$$\int \cos ax \; dx = \tfrac{1}{a}\sin ax + c$$

Don't forget the constant of integration.

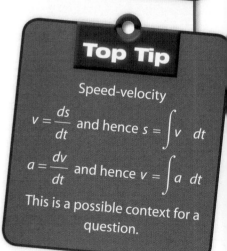

Top Tip

Speed-velocity

$$v = \frac{ds}{dt} \quad \text{and hence} \quad s = \int v \; dt$$

$$a = \frac{dv}{dt} \quad \text{and hence} \quad v = \int a \; dt$$

This is a possible context for a question.

Composite functions

When a function is composed of two functions and the 'inside' function is linear, we have a simple way of integrating it.

If $\int f(x)\ dx = F(x)$ then $\int f(ax+b)\ dx = \frac{1}{a}F(ax+b)$

Examples

(i) $\int \cos(5x)\ dx = \frac{1}{5}\sin(5x) + c$

(ii) $\int \sin(3x+2)\ dx = -\frac{1}{3}\cos(3x+2) + c$

(iii) $\int 5\cos(2x+1)\ dx = \frac{5}{2}\sin(2x+1) + c$

(iv) $\int (3x+1)^5\ dx = \frac{1}{3}\cdot\frac{(3x+1)^6}{6} + c = \frac{1}{18}(3x+1)^6 + c$

Problem solving example

(i) By using the fact that $\cos 2x = 1 - 2\sin^2 x$, find an expression for $\sin^2 x$.

(ii) Hence, or otherwise, find $\int \sin^2 x\ dx$

Response

(i) $\cos 2x = 1 - 2\sin^2 x$

$2\sin^2 x = 1 - \cos 2x$

$\Rightarrow \sin^2 x = \frac{1}{2} - \frac{1}{2}\cos 2x$

(ii) $\int \sin^2 x\ dx = \int \frac{1}{2} - \frac{1}{2}\cos 2x\ dx$

$= \frac{1}{2}x - \frac{1}{2}\cdot\frac{1}{2}\sin 2x + c$

$= \frac{1}{2}x - \frac{1}{4}\sin 2x + c$

This type of integral is a regular visitor to the exam.

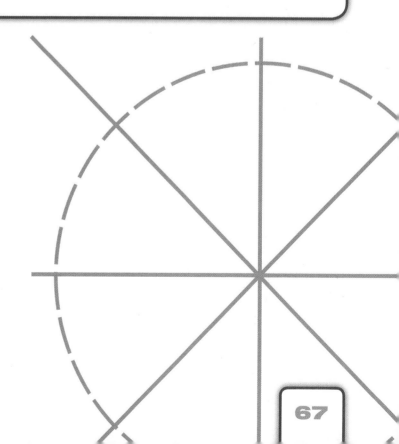

Logarithmic and Exponential Functions, part 1

You learned in Unit 1 that the log function and the exponential function are related – one is the inverse of the other.
In other words, one 'undoes' the other.
$\log_a(a^x) = x$ ... the log 'undoes' the exponential function.
$a^{\log_a x} = x$ ... the exponential 'undoes' the log function.

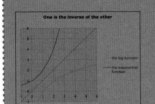
One is the inverse of the other
the log function
the exponential function

This leads to the most useful fact in this field:
$$y = a^x \Leftrightarrow \log_a y = \log_a(a^x) = x$$
i.e. $y = a^x \Leftrightarrow \log_a y = x$ ... remember this!

The arrow is double-headed. It will be useful to go in either direction in problems.

Top Tip

It is vital you can switch between log forms and exponential forms.
$y = a^x \Leftrightarrow \log_a y = x$
should be practised.

Logarithmic functions

Top Tip

Your calculator has a LOG button.
You can use it to work out the log of any number in any base, e.g.
$\log_2 5 = \text{LOG } 5 \div \text{LOG } 2$

Laws of logs
1   $\log_a 1 = 0$
2   $\log_a a = 1$
3   $\log_a(bc) = \log_a b + \log_a c$
4   $\log_a \left(\frac{b}{c}\right) = \log_a b - \log_a c$
5   $\log_a(b^n) = n\log_a b$
These rules can be used to simplify expressions.

Example 1
Simplify $\log_5 125 + \log_5 15$
Response
$\log_5 125 + \log_5 15$
$= \log_5(5^3) + \log_5(3.5)$ ... looking for powers or multiples of 5
$= 3\log_5 5 + \log_5 3 + \log_5 5$ ... using laws 5 and 3
$= 3 + \log_5 3 + 1$ ... using law 2
$= 4 + \log_5 3$

Example 2
Simplify $3\log_8 4 + \log_8 2 - {}^1/_2 \log_8 16$
Response

$3\log_8 4 + \log_8 2 - \frac{1}{2}\log_8 16$

$= \log_8\left(4^3\right) + \log_8 2 - \log_8\left(16^{\frac{1}{2}}\right)$

$= \log_8\left(\frac{4^3.2}{\sqrt{16}}\right)$

$= \log_8 32 = \log_8\left(2^5\right) = 5\log_8 2$

Now you should have the fact in your head that $8 = 2^3 \Rightarrow 2 = 8^{\frac{1}{3}}$

Thus $5\log_8 2 = 5\log_8\left(8^{\frac{1}{3}}\right) = \frac{1}{3}.5\log_8 8 = \frac{5}{3}$

Solving equations

In solving equations we make heavy use of the relation mentioned earlier: $y = a^x \Leftrightarrow \log_a y = x$
Note that in mathematics the most commonly used bases are 10 and e, where $e = 2.71828$ (to 6 s.f.).
Often we use $\ln x$ to represent $\log_e x$.

Logarithmic equations
Example 1
Solve $\ln x = 5$

Response
$\ln x = 5 \Leftrightarrow \log_e x = 5 \Leftrightarrow x = e^5 = 148.4$ (to 4 s.f.)

Example 2
Solve $\log_2 x = 1.6$

Response
$\log_2 x = 1.6 \Leftrightarrow x = 2^{1.6} = 3.031$ (to 4 s.f.)

Example 3
Solve $3\log_5 x + 1 = 8$

Response
$3\log_5 x + 1 = 8$
$\Rightarrow \log_5 x = {}^7/_3$
$\Leftrightarrow x = 5^{\frac{7}{3}} = 42.75$ (to 4 s.f.)

Problem solving example
The school bell rings and the thousand students start to go home.
The number of students, s, still in the school t minutes after the bell is modelled by
$s = 1000 - 600 \ln(t + 1)$. After how many minutes will the school be empty? Give your answer to 1 decimal place.

Response
We wish t when $s = 0$:
$1000 - 600 \ln(t + 1) = 0$
$\Rightarrow \ln(t + 1) = {}^{1000}/_{600}$
$\Rightarrow t + 1 = e^{\frac{1000}{600}} = 5.294...$
$\Rightarrow t = 4.3$ minutes (to 1 d.p.)

Exponential equations
Example 1
Solve $e^x = 3$

Response
$e^x = 3 \Leftrightarrow x = \log_e 3 = 1.099$ (to 4 s.f.)

Example 2
Solve $2^x = 1.5$

Response
$2^x = 1.5 \Leftrightarrow \ln(2^x) = \ln(1.5)$, taking the natural log of both sides.
$\Rightarrow x \ln 2 = \ln 1.5$
$\Rightarrow x = {}^{\ln 1.5}/_{\ln 2} = 0.5850$ (to 4 s.f.)

Example 3
Solve $3.4^x + 1 = 6$

Response
$3.4^x + 1 = 6 \Rightarrow 4^x = {}^5/_3 \Rightarrow \ln(4^x) = \ln({}^5/_3)$
$\Rightarrow x \ln(4) = \ln({}^5/_3)$
$\Rightarrow x = \ln({}^5/_3) / \ln(4) = 0.3685$ (to 4 s.f.)

Problem solving example
A rumour that there was going to be a fire-drill started spreading round the school. The spread of the rumour can be modelled by $p = 500e^{0.05t}$ where p pupils know the rumour after t minutes.
[$t = 0$ was when staff first became aware that a rumour was spreading.]
How long would it be, to the nearest minute, before 1000 pupils had heard the rumour?

Response
Here we wish to solve $500e^{0.05t} = 1000$
$500e^{0.05t} = 1000$
$\Rightarrow e^{0.05t} = \dfrac{1000}{500} = 2$
$\Rightarrow 0.05t = \ln(2)$
$\Rightarrow t = \dfrac{\ln(2)}{0.05} = 13.8629...$

It would be 14 minutes before 1000 pupils had heard the rumour.

Logarithmic and Exponential Functions, part 2

Some special equations

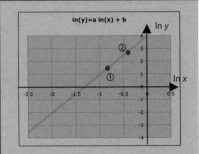

What follows is designated grade A/B material by the SQA.

Example 1
This graph has an equation of the form
$\ln(y) = a\ln(x) + b$ where a and b are constants.
When $x = 0.45$, $y = 5$.
When $x = 0.65$, $y = 15$.
Find the values of a and b correct
to the nearest whole number.

Response
Using the first pair of values:
$\ln(5) = a\ln(0.45) + b$
$\Rightarrow 1.61 = -0.80a + b$ ... ①

Using the second pair of values:
$\ln(15) = a\ln(0.65) + b$
$\Rightarrow 2.71 = -0.43a + b$ ... ②

Subtracting ② − ①: $1.1 = 0.37a$
$\Rightarrow a = 2.97$
Substitute in ①: $1.61 = -0.80.2.97 + b$
$\Rightarrow b = 1.61 + 0.80.2.97 = 3.986$
Thus $a = 3$ and $b = 4$ correct to the nearest whole numbers.
... and the graph is of $\ln(y) = 3\ln(x) + 4$.

Top Tip
Remember to answer the question. If the equation is asked for, don't just give the values of a and b.

Example 2
This graph has an equation of the form $y = ax^b$.
It passes through the points (9, 135) and (4, 40)
Find the actual equation.

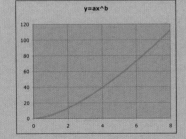

Response
It passes through (9, 135) $\Rightarrow 135 = a.9^b$ ... ①
It passes through (4, 40) $\Rightarrow 40 = a.4^b$ ... ②

$$① \div ②: \quad \frac{135}{40} = \frac{a.9^b}{a.4^b} \quad \Rightarrow \quad \frac{27}{8} = \left(\frac{9}{4}\right)^b$$

$$\Rightarrow \quad \ln\left(\frac{27}{8}\right) = \ln\left(\frac{9}{4}\right)^b = b\ln\left(\frac{9}{4}\right)$$

$$\Rightarrow \quad b = \ln\left(\frac{27}{8}\right) \div \ln\left(\frac{9}{4}\right) = 1.5$$

Substitute in ②: $40 = a.4^{1.5}$
$\Rightarrow a = 40 \div 4^{1.5} = 5$
... and the graph is of $y = 5x^{1.5}$

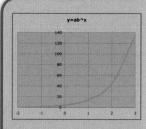

Example 3

This graph has an equation of the form $y = ab^x$. It passes through the points (1, 15) and (3, 135). Find the actual equation.

Response

It passes through (1, 15) $\Rightarrow$ $15 = a.b^1$ ... ①
It passes through (3, 135) $\Rightarrow$ $135 = a.b^3$ ... ②

$$② \div ① : \frac{135}{15} = \frac{a.b^3}{a.b^1}$$

$$\Rightarrow 9 = b^2 \Rightarrow b = 3$$

Substitute in ①: $15 = a.3^1 \Rightarrow a = 5$
... and the graph is of $y = 5.3^x$

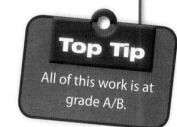

Top Tip

All of this work is at grade A/B.

Modelling

A scientist may carry out an experiment, gathering data.
He or she would then graph the data.

*   If the graph looked like Example 2 (note that it passes through the origin), he or she may try a model of the form $y = ax^b$ using two points on the curve to calculate a and b.
*   If the graph looked like Example 3 (note that it does not pass through the origin), he or she may try a model of the form $y = ab^x$ using two points on the curve to calculate a and b.

Example

A balloonist estimates the distance to the horizon, d km, using a formula of the form $d = ah^b$ where h is the height in metres and a and b are constants.
Experimental readings give: a distance of 1·6 km when the height is 16 m and a distance of 2 km when the height is 25 m
(i) Find a and b.
(ii) How far does the model predict the horizon will be at a height of 30 m?

Response

(i) $h = 16, d = 1·6$ $1·6 = a.16^b$ ... ①
     $h = 25, d = 2 \Rightarrow 2 = a.25^b$ ... ②

$$① \div ② : \Rightarrow \frac{2}{1·6} = \left(\frac{25}{16}\right)^b \Rightarrow \ln\left(\frac{2}{1·6}\right) = \ln\left(\frac{25}{16}\right)^b = b\ln\left(\frac{25}{16}\right)$$

$$\Rightarrow b = \ln\left(\frac{2}{1·6}\right) \div \ln\left(\frac{25}{16}\right) = 0·5$$

Substitute in ①: $1·6 = a.16^{0·5} = a.\sqrt{16} = 4a \Rightarrow a = 0·4$

(ii) The relationship is $d = 0·4h^{0·5}$ or $d = 0·4\sqrt{h}$
     When $h = 30, d = 0·4.\sqrt{30} = 2·19$ km

The Wave Function, part 1

Given the function $f(x) = a \cos x + b \sin x$, where a and b are constants, we can always express it in the forms:

$k \cos(x \pm a)$ or $k \sin(x \pm a)$ where k and a are positive constants.

A technique

Top Tip

It is common for students to cut corners in the exam and simply state the values of R and α. This will lose several communication marks.

Example

Express $3 \cos x + 4 \sin x$ in the form $k \sin(x + \alpha)$; $0 \leq \alpha \leq 2\pi$

Response

Step 1   Expand the desired form:

$k \sin(x + \alpha) = k \sin x \cos \alpha + k \cos x \sin \alpha$

Step 2   Compare this with the form you have:

Equate the coefficients of sin x: $4 = k \cos \alpha$ ... ①

Equate the coefficients of cos x: $3 = k \sin \alpha$ ... ②

Step 3   Square both equations and add them.

$4^2 + 3^2 = k^2 \cos^2\alpha + k^2 \sin^2\alpha$

$25 = k^2(\cos^2\alpha + \sin^2\alpha)$

$k^2 = 25$ ... remember $\cos^2\alpha + \sin^2\alpha = 1$

$\Rightarrow\ k = 5$ ... k is positive.

Step 4   Divide ② by ①   ... sine by cosine

$\dfrac{3}{4} = \dfrac{k \sin \alpha}{k \cos \alpha} = \tan \alpha$

$\Rightarrow \alpha = \tan^{-1}\left(\dfrac{3}{4}\right)$ or $\pi + \tan^{-1}\left(\dfrac{3}{4}\right)$ or $2\pi + \tan^{-1}\left(\dfrac{3}{4}\right)$

$\Rightarrow\ \alpha = 0 \cdot 644$ or $3 \cdot 785$ (The third answer is beyond the desired range.)

Step 5   Which of these angles should I take?

① tells us that cos is positive, so the angle is in the 1st or 4th quadrant.

② tells us that sin is positive, so the angle is in the 1st or 2nd quadrant.

Choose the option in the 1st quadrant ... $\alpha = 0 \cdot 644$

Step 6   Answer the question.

$3 \cos x + 4 \sin x = 5 \sin(x + 0 \cdot 644)$

Finding maxima and minima

Remember that $-1 \leq \sin A \leq 1$ for all A

So $\qquad -1 \leq \sin(x + \alpha) \leq 1$

... and $\qquad -k \leq k \sin(x + \alpha) \leq k$

A minimum, $-k$, will occur when the angle is $\frac{3\pi}{2}$ i.e. $x + \alpha = \frac{3\pi}{2}$

A maximum, k, will occur when the angle is $\frac{\pi}{2}$ i.e. $x + \alpha = \frac{\pi}{2}$

Further maxima and minima will occur at intervals of 2π.

It is a similar story with the cosine function:

$-k \leq k \cos(x + \alpha) \leq k$

A minimum, $-k$, will occur when the angle is π, i.e. $x + \alpha = \pi$.

A maximum, k, will occur when the angle is 0 or 2π, i.e. $x + \alpha = 0$ or 2α.

Further maxima and minima will occur at intervals of 2α.

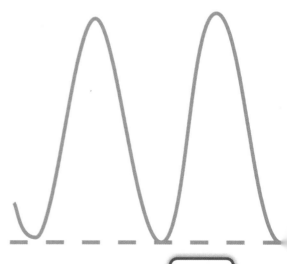

The Wave Function, part 2

Example 1

$f(x) = 5 \sin x° + 12 \cos x°$. Express $f(x)$ in the form $k \cos(x + α)°$; $0 \le α \le 360$.

State the minimum value of $f(x)$ and the value of x, $0 \le x \le 360$, at which it occurs.

Step 1 $k \cos(x + α)° = k \cos x° \cos α° - k \sin x° \sin α°$

Step 2 $12 = k \cos α°$ ... ①

             $5 = -k \sin α°$ $\Rightarrow$ $-5 = k \sin α°$ ... ②

Step 3 $k^2 = 5^2 + 12^2 = 169$

             $k = 13$

Step 4 $\tan α° = \dfrac{-5}{12}$

             $\Rightarrow α° = \tan^{-1}\left(\dfrac{-5}{12}\right)$ or $π + \tan^{-1}\left(\dfrac{-5}{12}\right)$ or $2π + \tan^{-1}\left(\dfrac{-5}{12}\right)$

             $\Rightarrow α = -22·6$ or $157·4$ or $337·4$

Step 5   ① tells us that cos is positive, so the angle is in the 1st or 4th quadrant.

             ② tells us that sin is negative, so the angle is in the 3rd or 4th quadrant.

             Choose the option in the 4th quadrant ... $α = 337·4$

Step 6 $f(x) = 13 \cos(x + 337·4)°$

             This will be at a minimum when $\cos(x + 337·4)° = -1$

             i.e. $f_{min} = -13$

             This will occur at $x + 337·4 = 180$ or 540 or 900 or ...

             $\Rightarrow x = -157·4$ or $202·6$ or $562·6$ or ...

             In the region $0 \le x \le 360$, the minimum is attained at $x = 202·6$

Top Tip

These can also be found using calculus but questions can always be posed to put this out of your reach.

Solving equations

Example

(i) Express $8 \cos x - 15 \sin x$ in the form $k \sin(x + \alpha)$; $0 \le \alpha \le 2\pi$.

(ii) Hence or otherwise solve the equation $8 \cos x - 15 \sin x + 10 = 0$, $0 \le x \le 2\pi$.

Response

Step 1 $k \sin(x + \alpha) = k \sin x \cos \alpha + k \cos x \sin \alpha$

Step 2 $-15 = k \cos \alpha$ ... ①

$\quad\quad\quad 8 = k \sin \alpha$ ... ②

Step 3 $k^2 = 15^2 + 8^2 = 289$

$\quad\quad\quad k = 17$

Step 4 $\tan \alpha = -\dfrac{8}{15}$

$\quad\quad\quad \Rightarrow \alpha = \tan^{-1}\left(\dfrac{-8}{15}\right)$ or $\pi + \tan^{-1}\left(\dfrac{-8}{15}\right)$ or $2\pi + \tan^{-1}\left(\dfrac{-8}{15}\right)$

$\quad\quad\quad \Rightarrow \alpha = -0{\cdot}490$ or $2{\cdot}65$ or $5{\cdot}79$

Step 5   ① tells us that cos is negative, so the angle is in the 2nd or 3rd quadrant.

② tells us that sin is positive, so the angle is in the 1st or 2nd quadrant. Choose the option in the 2nd quadrant ... $\alpha = 2{\cdot}65$

Step 6 $8 \cos x - 15 \sin x + 10 = 0$

$\quad\quad\quad \Rightarrow 17 \sin(x + 2{\cdot}65) + 10 = 0$

$\quad\quad\quad \Rightarrow \sin(x + 2{\cdot}65) = -{}^{10}/_{17}$

$\quad\quad\quad \Rightarrow x + 2{\cdot}65 = -0{\cdot}63, 3{\cdot}77, 5{\cdot}65, 10{\cdot}1\ldots$

$\quad\quad\quad \Rightarrow x = -3{\cdot}28, 1{\cdot}12, 3{\cdot}00, 7{\cdot}45\ldots$

In the region $0 \le x \le 2\pi$, $\Rightarrow x = 1{\cdot}12$ or $3{\cdot}00$

Top Tip

It is critical that you stay in the units used in the question ... degrees or radians.

This completes Maths 3.

For the unit test, you will be expected to show that you can perform the following under exam conditions. The SQA have published this list in their conditions and arrangements.

OUTCOME 1

Use vectors in three dimensions.

Performance criteria
a) Determine whether three points with given coordinates are collinear.
b) Determine the coordinates of the point which divides the join of two given points internally in a given numerical ratio.
c) Use the scalar product.

OUTCOME 2

Use further differentiation and integration.

Performance criteria
a) Differentiate $k \sin x, k \cos x$.
b) Differentiate using the function of a function rule.
c) Integrate functions of the form $f(x) = (x + q)^n, n$ rational except for -1, $f(x) = p \cos x$ and $f(x) = p \sin x$.

OUTCOME 3

Use properties of logarithmic and exponential functions.

Performance criteria
a) Simplify a numerical expression using the laws of logarithms.
b) Solve simple logarithmic and exponential equations.

OUTCOME 4

Apply further trigonometric relationships.

Performance criteria
a) Express $a \cos \theta + b \sin \theta$ in the form $r \cos (\theta \pm \alpha)$ or $r \sin (\theta \pm \alpha)$

Top Tip

Practice tests suitable for preparing for the unit test can be downloaded from **www.leckieandleckie.co.uk.**

Higher Mathematics

Paper 1

1 hour 30 minutes

Read carefully

Calculators may <u>NOT</u> be used in this paper.

Section A – Questions 1–20 (40 marks)

For this section of the examination you must use an **HB pencil**.

Section B (30 marks)

1 Full credit will be given only where the solution contains appropriate working.

2 Answers obtained by readings from scale drawings will not receive any credit.

Read carefully

1 Check that the answer sheet provided is for **Mathematics Higher (Section A)**

2 For this section of the examination you must use an **HB pencil** and, where necessary, an eraser.

3 Check that the answer sheet you have been given has **your name**, **date of birth**, **SCN** (Scottish Candidate Number) and **Centre Name** printed on it.

4 If any of this information is wrong, tell the Invigilator immediately.

5 If this information is correct, **print** your name and seat number in the boxes provided.

6 The answer to each question is **either** A, B, C or D. Decide what your answer is, then, using your pencil, put a horizontal line in the space provided (see sample question below).

7 There is **only one correct** answer to each question.

8 Rough working should **not** be done on your answer sheet.

9 At the end of the exam, put the **answer sheet for Section A inside the front cover of your answer book**.

Sample Question

A curve has equation $y = x^3 - 4x$.

What is the gradient at the point where $x = 2$?

    A   8

    B   1

    C   0

    D   −4

The correct answer is **A** — 8. The answer **A** has been clearly marked in **pencil** with a horizontal line (see below).

    A   **B**   **C**   **D**

Changing an answer

If you decide to change your answer, carefully erase your first answer and using your pencil, fill in the answer that you want. The answer below has been changed to **D**.

    A   **B**   **C**   **D**

FORMULAE LIST

Circle

The equation $x^2 + y^2 + 2gx + 2fy + c = 0$ represents a circle centre $(-g, -f)$ and radius $\sqrt{g^2 + f^2 - c}$.

The equation $(x - a)^2 + (y - b)^2 = r^2$ represents a circle centre (a, b) and radius r.

Scalar Product: $\boldsymbol{a.b} = |\boldsymbol{a}|\,|\boldsymbol{b}|\cos\theta$, where θ is the angle between $\boldsymbol{a}$ and $\boldsymbol{b}$

or $\boldsymbol{a.b} = a_1b_1 + a_2b_2 + a_3b_3$ where $\boldsymbol{a} = \begin{pmatrix} a_1 \\ a_2 \\ a_3 \end{pmatrix}$ and $\boldsymbol{b} = \begin{pmatrix} b_1 \\ b_2 \\ b_3 \end{pmatrix}$

Trigonometric formulae:

$$\sin(A \pm B) = \sin A \cos B \pm \cos A \sin B$$
$$\cos(A \pm B) = \cos A \cos B \mp \sin A \sin B$$
$$\sin 2A = 2\sin A \cos A$$
$$\cos 2A = \cos^2 A - \sin^2 A$$
$$= 2\cos^2 A - 1$$
$$= 1 - 2\sin^2 A$$

Table of standard derivatives:

$f(x)$	$f'(x)$
$\sin ax$	$a\cos ax$
$\cos ax$	$-a\sin ax$

Table of standard integrals:

$f(x)$	$\int f(x)\,dx$
$\sin ax$	$-\dfrac{1}{a}\cos ax + c$
$\cos ax$	$\dfrac{1}{a}\sin ax + c$

SECTION A

All questions should be attempted.

1. Calculate the gradient of the line *perpendicular* to the line with equation $2x + 5y = 1$.

   A $\dfrac{2}{5}$ B $-\dfrac{2}{5}$ C $\dfrac{5}{2}$ D $-\dfrac{5}{2}$

2. Here is a sketch of the graph $y = f(x)$.

   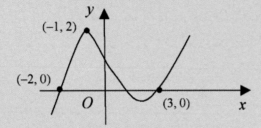

   Here is a sketch of the graph *of a related function*.

   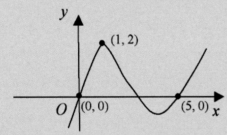

   Of which of the following is it most likely to be a sketch?

   A $y = f(x - 2)$ B $y = f(x + 2)$ C $y = f(x) - 2$ D $y = f(x) + 2$

3. What can be said about the recurrence relation
   $u_{n+1} = -\frac{2}{5}u_n + 210$ as n approaches infinity?

   A   It has a limit of 150

   B   It has a limit of 294

   C   It has a limit of 350

   D   It has a limit of 525

4. What is the equation of the line which passes through $(4, -1)$ and $(0, 7)$?

   A $y = 7 - 2x$ B $y = 2x + 7$ C $y = 7 - \frac{3}{2}x$ D $y = \frac{3}{2}x + 7$

5. Which of the following statements is true about $x^2 + y^2 - 8x + 4y - 5 = 0$?

    A   It represents a circle with centre $(-4, 2)$ and radius 5.

    B   It represents a circle with centre $(4, -2)$ and radius 5.

    C   It represents a circle with centre $(-8, 4)$ and radius 5.

    D   It does not represent a circle.

6. Express $3x^2 + 6x - 4$ in the form $a(x + b)^2 + c$.

    A $3(x + 2)^2 - 7$

    B $3(x + 1)^2 - 7$

    C $3(x - 2)^2 - 7$

    D $3(x - 1)^2 - 7$

7. What is the equation of a circle with centre $(-2, 5)$ having the x-axis as a tangent?

    A $(x - 2)^2 + (y + 5)^2 = 4$

    B $(x + 2)^2 + (y - 5)^2 = 4$

    C $(x - 2)^2 + (y + 5)^2 = 25$

    D $(x + 2)^2 + (y - 5)^2 = 25$

8. $f(x) = 2x - 3$ and $g(x) = x^2 + 2$

Find an expression for $g(f(x))$

    A $4x^2 - 12 + 11$

    B $4x^2 - 6x + 11$

    C $2x^2 + 1$

    D $2x^2 - 5$

9. What is the equation of the tangent to the curve $y = (3x + 1)^{\frac{1}{2}}$ at the point where $x = 1$?

    A $y - 2 = 3(x - 1)$

    B $y + 2 = 3(x - 1)$

    C $y - 2 = \frac{3}{4}(x - 1)$

    D $y + 2 = \frac{3}{4}(x - 1)$

10. The vectors $4\boldsymbol{i} + \boldsymbol{j} - 2\boldsymbol{k}$ and $x\boldsymbol{i} + 2\boldsymbol{j} + x\boldsymbol{k}$ are mutually perpendicular. What is the value of x?

    A -1

    B -2

    C   1

    D   2

11. Differentiate $\dfrac{x^2 - x}{\sqrt{x}}$

    A $\frac{3}{2}x^{-\frac{1}{2}} - \frac{1}{2}x^{\frac{1}{2}}$

    B $\frac{3}{2}x^{\frac{1}{2}} - \frac{1}{2}x^{-\frac{1}{2}}$

    C $\frac{1}{2}x^{\frac{1}{2}} - \frac{1}{2}x^{-\frac{1}{2}}$

    D $\frac{2}{5}x^{\frac{5}{2}} - \frac{2}{3}x^{\frac{3}{2}}$

12. What is $\log(xy^3)$ equal to?

    A $\log x + 3\log y$

    B $3(\log x + \log y)$

    C $3 \log x \log y$

    D $\log x + \log y + \log 3$

13. Given that $\tan x^\circ = \frac{5}{12}$ and that $0 < x < 90$,
what is the exact value of $\cos 2x^\circ$?

    A 1

    B $\frac{10}{13}$

    C $\frac{119}{169}$

    D $\frac{120}{169}$

14. The points P(4, 2, −2), Q(5, y, z) and R(7, −1, 4) are collinear.
What are the values of y and z?

    A $y = -1;\ z = 2$

    B $y = 1;\ z = -2$

    C $y = -1;\ z = 0$

    D $y = 1;\ z = 0$

15. What must be true about m so that the equation
$y = 2x^2 + mx + 1$, $m > 0$ has real, distinct roots?

    A $m > 2\sqrt{2}$

    B $m \geq 2\sqrt{2}$

    C $m < 2\sqrt{2}$

    D $m \leq 2\sqrt{2}$

16. $75 = 45 + 30$. What is the **exact** value of $\cos 75^\circ$?

    A $\dfrac{1 + \sqrt{3}}{2\sqrt{2}}$

    B $\dfrac{1 - \sqrt{3}}{2\sqrt{2}}$

    C $\dfrac{\sqrt{3} - 1}{2\sqrt{2}}$

    D $\dfrac{\sqrt{3} + 1}{2\sqrt{2}}$

17. Which one of the options is a factor of $x^3 - 6x^2 + 11x - 6$?

    A $(x + 1)$ B $(x - 2)$ C $(x + 3)$ D $(x - 4)$

18. Integrate $\dfrac{3x^3 + 2x}{x}$

    A $\dfrac{\frac{3}{4}x^4 + x^2}{\frac{1}{2}x^2} + c$

    B $9x^2 + 2 + c$

    C $3x^2 + 2 + c$

    D $x^3 + 2x + c$

19. $\displaystyle\int_{\frac{\pi}{6}}^{\frac{\pi}{3}} \cos 3x \; dx$ is equal to:

    A 1

    B -1

    C $\frac{1}{3}$

    D $-\frac{1}{3}$

20. In which quadrant does a lie when
$3 \sin x - 4 \cos x$ is expressed in the form $r \sin(x + a)$?

    A 1st quadrant

    B 2nd quadrant

    C 3rd quadrant

    D 4th quadrant

[END OF SECTION A]

SECTION B

All Questions should be attempted.

21. The movement of a particular chair on a ferris wheel can be modelled by the function $h(x) = 2\cos x° + 2\sqrt{3} \sin x° + 5$ where h is the height in metres and $x°$ is the angle turned through since the start of the ride.

Marks

(a) Express $2\cos x° + 2\sqrt{3} \sin x°$ in the form $k \cos(x - a)°$ where $k > 0$ and $0 \le a \le 360$. **4**

(b) State the minimum and maximum height of the chair. **1**

(c) What is the value of x when this minimum occurs in the first turn of the wheel?
[Working must be shown.] **2**

22. A function is defined by $f(x) = x^3 - 12x^2 + 45x - 54$.

(a) Show that $x = 3$ is a root to the equation $f(x) = 0$. **1**

(b) Completely factorise $f(x)$. **3**

(c) Find the stationary points on the curve $y = f(x)$ and determine their nature. **6**

23. The diagram shows the circle with equation $x^2 + y^2 = 25$. Tangents AT and BT intersect at T(13, 0).

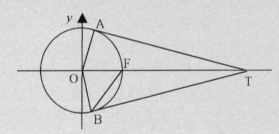

Find the exact value of

(a) $\sin \angle AOT$ **2**

(b) $\sin \angle AOB$ **2**

(c) $\cos \angle FBT$ **3**

24. A recurrence relation takes the form $u_{n+1} = k^2 u_n + 3k$ where k is a constant.

(a) Write down expressions for the first three terms if $u_0 = 0$. **2**

(b) The recurrence relation has a limit of 2.
Find the value of k. **4**

[END OF SECTION B]

[END OF QUESTION PAPER]

Higher Mathematics

Paper 2

1 hour 10 minutes

Read Carefully

1 **Calculators may be used in this paper.**

2 Full credit will be given only where the solution contains appropriate working.

3 Answers obtained by readings from scale drawings will not receive any credit.

FORMULAE LIST

Circle

The equation $x^2 + y^2 + 2gx + 2fy + c = 0$ represents a circle centre $(-g, -f)$ and radius $\sqrt{g^2 + f^2 - c}$.

The equation $(x - a)^2 + (y - b)^2 = r^2$ represents a circle centre (a, b) and radius r.

Scalar Product: $\quad a.b = |a|\,|b|\cos\theta$, where θ is the angle between a and b

or $\quad a.b = a_1b_1 + a_2b_2 + a_3b_3$ where $a = \begin{pmatrix} a_1 \\ a_2 \\ a_3 \end{pmatrix}$ and $b = \begin{pmatrix} b_1 \\ b_2 \\ b_3 \end{pmatrix}$

Trigonometric formulae:
$$\sin(A \pm B) = \sin A \cos B \pm \cos A \sin B$$
$$\cos(A \pm B) = \cos A \cos B \mp \sin A \sin B$$
$$\sin 2A = 2\sin A \cos A$$
$$\cos 2A = \cos^2 A - \sin^2 A$$
$$= 2\cos^2 A - 1$$
$$= 1 - 2\sin^2 A$$

Table of standard derivatives:

$f(x)$	$f'(x)$
$\sin ax$	$a\cos ax$
$\cos ax$	$-a\sin ax$

Table of standard integrals:

$f(x)$	$\int f(x)\,dx$
$\sin ax$	$-\dfrac{1}{a}\cos ax + c$
$\cos ax$	$\dfrac{1}{a}\sin ax + c$

All questions should be attempted.

1. ABC is a triangle with vertices A(–5, 11), B(5, 1) and C(9, 3).

    (*a*) Find the equation of the perpendicular bisector of BC.   4

    (*b*) Find the equation of the median CM where M lies on AB.   2

    (*c*) Find the coordinates of K, the point where these two lines intersect.   3

    (*d*) In what ratio does K divide CM?   2

2. Relative to a suitable set of axes and using convenient units, three hills have peaks at A(7, 9, 12), B(–2, –3, 5) and C(–1, 4, 15).

    (*a*) How far is it from B to C as the crow flies?   2

    (*b*) A hill walker on A is looking at B. He turns to look at C. Through what angle did his head turn?   6

3. (*a*) Show that $x^2 + y^2 - 2x - 2y - 11 = 0$ represents a circle.   1

    (*b*) Find the coordinates of the points where the line $x - 5y = 9$ cuts this circle.   4

4. A manufacturer makes small tins from a sheet of metal which is 10 cm by 20 cm. Squares of edge x cm are cut from it as shown and the edges are bent up to form two halves of the box.

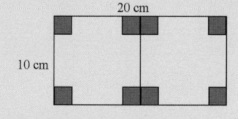 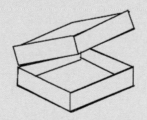 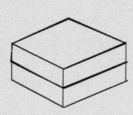

    (*a*) Show that the volume of the box is given by $V = 8x(5 - x)^2$ cm^3.   2

    (*b*) Find the maximum volume and the value of x when this occurs.   6

5. (*a*) Show that $f(x) = 4x + \sin 3x$ is an increasing function.   3

    (*b*) Find the equation of the tangent to the curve $y = f(x)$ when $x = \frac{2\pi}{3}$.   3

6. (a) (i) Show that $(x - 1)$ is a factor of $x^3 - 2x^2 - 19x + 20$.   1

(ii) Hence factorise the expression completely.   3

(b) Calculate the area trapped between the curves
$y = x^3 - 4x^2 - 7x + 9$ and $y = -2x^2 + 12x - 11$ 5

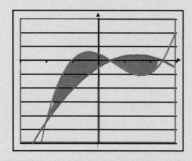

7. Solve the equation $2\cos 2x° + \sin x° + 1 = 0, \ 0 \le x \le 360$ 5

8. The rate at which water flows over a waterfall,
y litres/min, is related to the head of water, x metres.
The relation takes the form $y = ax^b$ where a and b are constants.

(a) Show that the relation between $\log x$ and $\log y$ is linear.

(b) The table shows some data that has been collected about
a couple of waterfalls.
Use this information to find the values of a and b
and hence the relation between x and y.

Head of water (x metres)	Rate of flow (y litres/min)	$\log_{10}$(head)	$\log_{10}$(rate)
5	83·85	0·7	1·92
35	10870·8	1·54	4·04

    3

    5

[END OF QUESTION PAPER]

Marking Scheme, Paper 1

Question	Answer	Comment
1	**Option C** rearrange given line to get $y = -\frac{2}{5}x + \frac{1}{5}$ $m_{\text{line}} = -\frac{2}{5} \Rightarrow m_{\text{perp}} = \frac{5}{2}$	2 marks or zero
2	**Option A** by inspection ... a translation of 2 in the x-direction.	2 marks or zero
3	**Option A** multiplier a proper fraction $\Rightarrow$ limit $L = -\frac{2}{5}L + 210 \Rightarrow L = 150$	2 marks or zero
4	**Option A** $m = (7 + 1)/(0 - 4) = -2$	2 marks or zero
5	**Option B** by inspection centre $(4, -2)$ so radius $= \sqrt{(4^2 + 2^2 - (-5))} = 5$	2 marks or zero
6	**Option B** $3(x^2 + 2) - 4 = 3[(x + 1)^2 - 1] - 4$ $= 3(x + 1)^2 - 7$	2 marks or zero
7	**Option D** x axis as tangent $\Rightarrow$ radius 5	2 marks or zero
8	**Option A** $(2x - 3)^2 + 2$	2 marks or zero
9	**Option C** $y_{x=1} = (3 + 1)^{\frac{1}{2}}$; $\frac{dy}{dx}_{(x=1)} = \frac{1}{2}(3x + 1)^{-\frac{1}{2}}.3 = \frac{3}{4}$	2 marks or zero
10	**Option A** scalar product $= 0 \Rightarrow 4x + 2 - 2x = 0 \Rightarrow x = -1$	2 marks or zero
11	**Option B** $\frac{d}{dx}(x^{\frac{3}{2}} - x^{\frac{1}{2}})$	2 marks or zero
12	**Option A** $\log x + \log y^3 = \log x + 3\log y$	2 marks or zero
13	**Option C** $\cos 2x = \cos^2 x - \sin^2 x = \frac{144}{169} - \frac{25}{169}$	2 marks or zero
14	**Option D** $\boldsymbol{r} - \boldsymbol{p} = 3\boldsymbol{i} - 3\boldsymbol{j} + 6\boldsymbol{k}$; $\boldsymbol{q} - \boldsymbol{p} = \boldsymbol{i} + (y - 2)\boldsymbol{j} + (z + 2)\boldsymbol{k}$ $3(\boldsymbol{q} - \boldsymbol{p}) = 3\boldsymbol{i} + 3(y - 2)\boldsymbol{j} + 3(z + 2)\boldsymbol{k}$ equate components	2 marks or zero

15		**Option A** $b^2 - 4ac > 0$ ... strictly bigger than	2 marks or zero
16		**Option C** cos 75 = cos 45 cos 30 − sin 45 sin 30 = ½.√3⁄2 − ½.½	2 marks or zero
17		**Option B** $f(2) = 0$	2 marks or zero
18		**Option D** $f(x) = 3x^2 + 2$	2 marks or zero
19		**Option D** = ⅓ sinπ − ⅓ sin½	2 marks or zero
20		**Option D** $3 = r \cos a$ (1st or 4th); $-4 = r \sin a$ (3rd or 4th)	2 marks or zero
21	(a)	**Answer: 4 cos(x − 60)°** $2\cos x° + 2\sqrt{3} \sin x° = k \cos(x − a)°$ $= k \cos x° \cos a° + k \sin x° \sin a°$ $\Rightarrow 2 = k \cos a°$ (1st or 4th quadtant) $\Rightarrow 2\sqrt{3} = k \sin a°$ (1st or 2nd quadrant) $\tan a° = \sqrt{3} \Rightarrow a = 60$ (1st quadrant). $k^2 \cos^2 a° + k^2 \sin^2 a° = 4 + 12 = 16$ $\Rightarrow k^2 (\cos^2 a° + \sin^2 a°) = k^2 = 16$ $\Rightarrow k = 4$ $4 \cos(x − 60)°$	•1 expansion (explicit) •2 equate coeffs (explicit) •3 process to get a •4 process to get k
	(b)	**Answer: max = 9; min = 1** $h(x)_{max} = 4 + 5 = 9$ $h(x)_{min} = -4 + 5 = 1$	•5 process for *min and max*
	(c)	**Answer: x = 240** For min $\cos(x − 60)° = -1$ $\Rightarrow x − 60 = 180$ $\Rightarrow x = 240$	•6 condition for minimum •7 x value at minimum

22	(a)	**Answer: *proof***
		 3 | 1 −12 45 −54
		      3 −27 54
		 ───────────
		     1 −9 18 0
			•1 get 0 and conclusion

	(b)	**Answer: $f(x) = (x − 3)(x − 3)(x − 6)$** remainder zero $\Rightarrow x − 3$ is a factor $\Rightarrow x = 3$ is a root $f(x) = (x − 3)(x^2 − 9x + 18)$ $\Rightarrow f(x) = (x − 3)(x − 3)(x − 6)$	•2 quadratic factor •3 quad factorised •4 $f(x)$ factorised (explicit)

(c)	**Answer: max TP at (3, 0); min TP at (5, –4)**	$\bullet^5$	know to differentiate
	$f'(x) = 3x^2 - 24x + 45$	$\bullet^6$	differentiate
	$f'(x) = 0$ at stationary points	$\bullet^7$	equate to zero (explicit)
	$3x^2 - 24x + 45 = 0$	$\bullet^8$	x–values
	$\Rightarrow x^2 - 8x + 15 = (x - 3)(x - 5) = 0$	$\bullet^9$	table of signs
	$\Rightarrow x = 3$ or $x = 5$	$\bullet^{10}$	nature and y-values
	$\Rightarrow y = 0$ or $y = -4$		

Table of signs:

x	$\rightarrow$	3	$\rightarrow$	5	$\rightarrow$
$(x - 3)$	–	0	+	+	+
$(x - 5)$	–	–	–	0	+
dy/dx	+	0	–	0	+
gradient	/	—	\	—	/
		max		min	

max TP at (3, 0); min TP at (5, –4)

23 (a)	**Answer: ¹²⁄₁₃**	$\bullet^1$	find sizes (interpret diagram)
	OT = 13; OA = 5 $\angle$OAT = 90° $\Rightarrow$ AT = 12	$\bullet^2$	find value
	$\Rightarrow \sin \angle$AOT = ¹²⁄₁₃		
(b)	**Answer: ¹²⁰⁄₁₆₉**		
	$\sin\angle$AOB = $\sin 2\angle$AOT =		
	$2\sin\angle$AOT $\cos\angle$AOT		
	$= 2. $ ¹²⁄₁₃. ⁵⁄₁₃ $= $ ¹²⁰⁄₁₆₉	$\bullet^3$	double angle
		$\bullet^4$	evaluate
(c)	**Answer: ³⁄₁₃**		
	$2\angle$OBF = 180° – $\angle$AOT		
	$\Rightarrow \angle$OBF = 90° – $\angle$AOT/2		
	$\angle$FBT = 90° – $\angle$OBF = $\angle$AOT/2	$\bullet^5$	recognise half-angle
	$\cos 2x = 2\cos^2 x - 1$	$\bullet^6$	express $\cos x$ in terms of $\cos 2x$
	If $2x = \angle$AOT then $x = \angle$FBT	$\bullet^7$	evaluate
	$\cos 2x = 2\cos^2 x - 1$		

$$\Rightarrow \cos x = \sqrt{\frac{\cos 2x + 1}{2}}$$

$$\Rightarrow \cos\angle\text{FBA} = \sqrt{\frac{⁵⁄₁₃ + 1}{2}} = \sqrt{\frac{5 + 13}{26}} = \sqrt{\frac{18}{26}} = \sqrt{\frac{9}{13}}$$

24 (a)	**Answer: $3k$, $3k^3 + 3k$, $3k^5 + 3k^3 + 3k$**		
	$u_0 = 0$	$\bullet^1$	interpret (u_1 and u_2)
	$u_1 = k^2.0 + 3k = 3k$	$\bullet^2$	process u_3
	$u_2 = k^2.(3k) + 3k = 3k^3 + 3k$		
	$u_3 = k^2.(3k^3 + 3k) + 3k = 3k^5 + 3k^3 + 3k$		

(b)	**Answer: $k = \frac{1}{2}$** limit $= 2$ $\Rightarrow 2 = 2k^2 + 3k$ $\Rightarrow 2k^2 + 3k - 2 = 0$ $\Rightarrow (2k - 1)(k + 2) = 0$ $\Rightarrow k = \frac{1}{2}$ or -2 however for a limit to exist, $k^2 < 1$ $\Rightarrow k = \frac{1}{2}$	$\bullet^3$ strategy for finding limit $\bullet^4$ form quadratic equation $\bullet^5$ solve for 2 values $\bullet^6$ select value with reason

Marking Scheme, Paper 2

Question	Answer	Comment
1 (a)	**Answer: $2x + y = 16$ or *equivalent*** midpoint BC = (7, 2) $m_{BC} = {}^{(3-1)}\!/_{(9-5)} = \frac{1}{2}$ $\Rightarrow m_{perp} = -2$ equation of perp: $y - 2 = -2(x - 7)$	•^{1} strategy (find midpoint) •^{2} find gradient •^{3} find perp •^{4} equation
(b)	**Answer: $x + 3y = 18$ or *equivalent*** M is (0, 6) $m_{MC} = {}^{(6-3)}\!/_{(0-9)} = -\frac{1}{3}$ equation of median: $y - 6 = -\frac{1}{3}x$	•^{5} find gradient and midpoint •^{6} find equation
(c)	**Answer: K(6, 4)** from (b) $x = 18 - 3y$ substitute in (a): $2(18 - 3y) + y = 16$ $\Rightarrow 36 - 6y + y = 16$ $\Rightarrow 5y = 20$ $\Rightarrow y = 4$ $\Rightarrow x = 6$ $\Rightarrow$ K(6, 4)	•^{7} strategy: form equation •^{8} solve for x •^{9} solve for y and report
(d)	**Answer: 1:2 or $\frac{1}{2}$** $\dfrac{\vec{CK}}{\vec{KM}} = \dfrac{k-c}{m-k} = \dfrac{\begin{bmatrix}6\\4\end{bmatrix} - \begin{bmatrix}9\\3\end{bmatrix}}{\begin{bmatrix}0\\6\end{bmatrix} - \begin{bmatrix}6\\4\end{bmatrix}} = \dfrac{\begin{bmatrix}-3\\1\end{bmatrix}}{\begin{bmatrix}-6\\2\end{bmatrix}} = \dfrac{\begin{bmatrix}-3\\1\end{bmatrix}}{2\begin{bmatrix}-3\\1\end{bmatrix}} = \dfrac{1}{2}$	•^{10} strategy •^{11} process Note: vectors are not essential … a simple step-on process is possible
2 (a)	**Answer $5\sqrt{6}$ units** $\vec{BC} = c - b = \begin{bmatrix}-1\\4\\15\end{bmatrix} - \begin{bmatrix}-2\\-3\\5\end{bmatrix} = \begin{bmatrix}1\\7\\10\end{bmatrix}$ $|\vec{BC}| = \sqrt{1 + 49 + 100} = 5\sqrt{6}$	•^{1} strategy (e.g. distance formula) •^{2} process
(b)	**Answer $47 \cdot 4°$** $\vec{AB} = b - a = \begin{bmatrix}-2\\-3\\5\end{bmatrix} - \begin{bmatrix}7\\9\\12\end{bmatrix} = \begin{bmatrix}-9\\-12\\-7\end{bmatrix}$ $|\vec{AB}| = \sqrt{81 + 144 + 49} = \sqrt{274}$ $\vec{AC} = c - a = \begin{bmatrix}-1\\4\\15\end{bmatrix} - \begin{bmatrix}7\\9\\12\end{bmatrix} = \begin{bmatrix}-8\\-5\\3\end{bmatrix}$	•^{3} find directed lines AB and AC •^{4} correct strategy i.e. formula for cos $\angle$BAC with substitution

$|\overrightarrow{AC}| = \sqrt{64 + 25 + 9} = \sqrt{98}$

$\overrightarrow{AB}.\overrightarrow{AC} = \begin{bmatrix} -9 \\ -12 \\ -7 \end{bmatrix}\begin{bmatrix} -8 \\ -5 \\ 3 \end{bmatrix} = 72 + 60 - 21 = 111$

$\cos\angle BAC = \dfrac{111}{\sqrt{274 \times 98}} = 0.677$

$\Rightarrow \angle BAC = 47.4°$

- $\bullet^5$ interpret scalar product
- $\bullet^6$ process magnitude AB
- $\bullet^7$ process magnitude AC
- $\bullet^8$ find $\angle BAC$

---

3 (a)

Answer: proof $g^2 + f^2 - c > 0$

$x^2 + y^2 - 2x - 2y - 11 = 0$

$\Rightarrow -g = 1, -f = 1, c = -11$

$g^2 + f^2 - c = 1 + 1 + 11 = 13 > 0$

- $\bullet^1$ interpret equation

(b)

Answer: (–1, –2) and (4, –1)

substitute $x = 9 + 5y$ into circle:

$(9 + 5y)^2 + y^2 - 2(9 + 5y) - 2y - 11 = 0$

$\Rightarrow 26y^2 + 78y + 52 = 0$

$\Rightarrow y^2 + 3y + 2 = 0$

$\Rightarrow (y + 2)(y + 1) = 0$

$\Rightarrow y = -2, -1$

$\Rightarrow x = -1, 4$

points are (–1, –2) and (4, –1)

- $\bullet^2$ strategy: substitution
- $\bullet^3$ process: get quadratic
- $\bullet^4$ solve for x
- $\bullet^5$ solve for y and report

---

4 (a)

Answer: *proof*

$V = lbh$

$\Rightarrow V = 2x \times (10 - 2x) \times (10 - 2x)$

$\Rightarrow V = 2x.2(5 - x).2.(5 - x)$

$\Rightarrow = 8x(5 - x)^2$

- $\bullet^1$ strategy: start proof
- $\bullet^2$ process: to conclusion

(b)

$V = 8x(5 - x)^2 = 200x - 80x^2 + 8x^3$

$\frac{dV}{dx} = 200 - 160x + 24x^2$

$\frac{dV}{dx} = 0$ at stationary points.

$200 - 160x + 24x^2 = 0$

$\Rightarrow 3x^2 - 20x + 25 = 0$

$\Rightarrow (3x - 5)(x - 5) = 0$

$\Rightarrow x = \tfrac{5}{3}$ or 5

- $\bullet^3$ know to differentiate
- $\bullet^4$ differentiate
- $\bullet^5$ equate to zero for SPs
- $\bullet^6$ solve quadratic
- $\bullet^7$ table of signs to eliminate the 'min' value
- $\bullet^8$ evaluate V max

x	$\rightarrow$	$\tfrac{5}{3}$	$\rightarrow$	5	$\rightarrow$
$3x - 5$	–	0	+	+	+
$x - 5$	–	–	–	0	+
$\frac{dy}{dx}$	+	0	–	0	+
	/	—	\	—	/
		max		min	

maximum occurs when $x = \tfrac{5}{3}$

$\Rightarrow V_{\max} = 8 \times \tfrac{5}{3} \times (5 - \tfrac{5}{3}) = 148$ cm^3

(to nearest whole number)

5	(a)	**Answer:** $dy/dx > 0$ $f'(x) = 4 + 3\cos 3x$ $-1 \le \cos 3x \le 1$ $\Rightarrow -3 \le 3\cos 3x \le 3$ $\Rightarrow 1 \le 4 + 3\cos 3x \le 7$ $\Rightarrow f'(x) > 0$ for all x $\Rightarrow f(x)$ is an increasing function	•^{1} strategy: know condition for increasing function •^{2} communicate: restriction on cos function •^{3} process: to conclusion
	(b)	**Answer:** $y - x = \pi$ $f(\frac{\pi}{3}) = 4 \cdot \frac{\pi}{3} + \sin \pi = \frac{4\pi}{3}$ $f'(\frac{\pi}{3}) = 4 + 3\cos \pi = 4 - 3 = 1$ equation of tangent: $y - \frac{4\pi}{3} = 1(x - \frac{\pi}{3})$ $\Rightarrow y - x = \pi$	•^{4} evaluate $f(\frac{\pi}{3})$ for point •^{5} evaluate $f'(\frac{\pi}{3})$ for gradient •^{6} communicate equation of line
6	(a)	**Answer:(i) Proof (ii)** $(x - 1)(x - 5)(x + 4)$ (i) $\begin{array}{c|cccc} 1 & 1 & -2 & -19 & 20 \\ & & 1 & -1 & -20 \\ \hline & 1 & -1 & -20 & 0 \end{array}$ remainder = zero $\Rightarrow x - 1$ is a factor (ii) it follows that $x^2 - x - 20$ is a factor $\Rightarrow (x - 5)(x + 4)$ are factors $\Rightarrow f(x) = (x - 1)(x - 5)(x + 4)$	•^{1} strategy: division with zero remainder •^{2} establish quadratic factor •^{3} process: factorise quadratic •^{4} communicate factorised form
	(b)	**Answer: 121 units^2** curves intersect where $x^3 - 4x^2 - 7x + 9 = -2x^2 + 12x - 11$ $\Rightarrow x^3 - 2x^2 - 19x + 20 = 0$ $\Rightarrow (x - 1)(x - 5)(x + 4) = 0$ $\Rightarrow x = 1, 5, -4$ area comes in two parts: $A = \int_{-4}^{1} x^3 - 2x^2 - 19x + 20 \ dx$ $= \left[\dfrac{x^4}{4} - \dfrac{2x^3}{3} - \dfrac{19x^2}{2} + \dfrac{20x}{1}\right]_{-4}^{1} = \dfrac{121}{12} + \dfrac{1504}{12}$ $= 135\frac{5}{12}$ $B = \int_{1}^{5} x^3 - 2x^2 - 19x + 20 \ dx$ $= \left[\dfrac{x^4}{4} - \dfrac{2x^3}{3} - \dfrac{19x^2}{2} + \dfrac{20x}{1}\right]_{1}^{5} = -\dfrac{775}{12} - \dfrac{121}{12}$ $= -74\frac{8}{12}$ the area is the magnitude of this integral ... $74\frac{8}{12}$ units^2 total area = 210 units^2 to nearest whole no.	•^{5} strategy: identify two areas to add •^{6} know to integrate (upper factor – lower) •^{7} process: integrate •^{8} process: deal with limits •^{9} deal with 'negative'

7	**Answer: $x = 90$ or $228\cdot6$** $2\cos 2x° + \sin x° + 1 = 0$ $\Rightarrow 2(1 - 2\sin^2 x°) + \sin x° + 1 = 0$ $\Rightarrow 4\sin^2 x° - \sin x° - 3 = 0$ $\Rightarrow \sin x° = 1$ or $-\frac{3}{4}$ $\Rightarrow x = 90, -48\cdot6, 228\cdot6$ $\Rightarrow x = 90$ or $228\cdot6$ or $(360 + (-48\cdot6)) = 311\cdot4$	•¹ strategy: deal with double angle •² form quadratic •³ solve for $\sin x$ •⁴ solve for 1 value of x •⁵ solve for all x in range... no more or less
8 (a)	**Answer: *proof*** $y = ax^b$ $\Rightarrow \log y = \log(ax^b)$ $\Rightarrow \log y = \log(a) + \log(x^b)$ $\Rightarrow \log y = \log(a) + b\log(x)$ if you plot $\log y$ against $\log x$ you get a straight line gradient b and y-intercept $\log a$	•¹ strategy: take logs •² apply laws of logs •³ identify linear model correctly
(b)	**Answer: $y = 1\cdot5x^{2\cdot5}$** using $(0\cdot7, 1\cdot92)$: $1\cdot92 = 0\cdot7b + \log a$ using $(1\cdot54, 4\cdot04)$: $4\cdot04 = 1\cdot54b + \log a$ subtracting: $2\cdot12 = 0\cdot84b$ $\qquad \Rightarrow b = 2\cdot5$ substitute: $1\cdot92 = 0\cdot7 \times 2\cdot5 + \log a$ $\qquad \Rightarrow \log a = 0\cdot17 \Rightarrow a = 10^{0\cdot17}$ $\qquad \Rightarrow a = 1\cdot5$ so model is $y = 1\cdot5x^{2\cdot5}$	•⁴ strategy: substitute to get system of equations •⁵ subtract to establish value of b •⁶ substitute to establish value of $\log a$ •⁷ establish value of a •⁸ communicate model